W9-AWW-481

PRAISE FOR *WHEN GADGETS BETRAY US*:

"[Robert Vamosi] points the finger at our unthinking relationship with technology. We put our faith in gizmos . . . [but they] are riddled with glitches and security flaws. And it's not just navigation devices: Vamosi takes us on a tour of digital cock-ups involving everything from mobile phones to wireless keyboards. . . . The interplay between humans and their gadgets is fascinating and complex. It is shaped by economics and psychology and the cultures we live in. Somewhere in the mix of those forces there may be a recipe for a more judicious use of technology, for some blend of techno-enthusiasm and common sense." —*New Scientist*

"*When Gadgets Betray Us*, Robert Vamosi's meticulously researched new book, offers a revealing look at the dark underbelly of our rapidly advancing electronics. This is not some Orwellian indictment of new technology, but instead a call for caution: Our gadgets are evolving faster than we can successfully secure them." —Salon.com

"A fascinating overview of 'hardware hacking,' from lockpicking and stealing cars to tapping mobile phones or cloning Oyster cards and passports. The vulnerabilities in modern tech that Vamosi describes can be alarming. . . . This text itself could, of course, make a fine mischief-maker's cookbook." —*Guardian* (UK)

"*When Gadgets Betray Us* helps us become aware of the benefits and the shortfalls of many scientific marvels." —*Post and Courier*

"This book isn't a Luddite call to smash our smart phones. Vamosi is careful to point out how mobile tech is helping the human race worldwide, as well. All he does is ask that every sexy new gizmo be greeted with a healthy dose of skepticism, and that we follow a few basic rules that will leave us with little to fear." —*BBC Focus*, four-star review

"[Vamosi's] message: We are so dazzled by our bright, shiny tech toys that we continue to strike the wrong balance between convenience and security. We must be more aware of the risks we are taking and learn to be more vigilant." —*San Jose Mercury News*

"[A] sobering, sometimes frightening book." —*NJ Star Ledger*

"Important." —*Good Men Project Magazine*

BUSINESS/SCIENCE/TECHNOLOGY DIVISION
CHICAGO PUBLIC LIBRARY
400 SOUTH STATE STREET
CHICAGO, IL 60605

"A solid analysis of just how deeply technology can be used to gather personal information about us without our awareness, a scenario more alarming than we can imagine." —*Publishers Weekly*

"A compelling scrutiny of the ways in which technological enhancements can be exploited for nefarious purposes. . . . An erudite wake-up call."
—*Kirkus Reviews*

"[Robert Vamosi] exposes a technology-development landscape chock-full of inadequately guarded data and programming. . . . Read this, and you'll never again ignore the default security settings on accounts or your devices again."
—*Library Journal*

"Vamosi is a skilled writer and the topic is fascinating." —*Law Technology News*

"*When Gadgets Betray Us* will make you think twice about many daily activities and the free wi-fi at your favorite coffee shop."
—Whiz BANG!, Scientopia Network

"You'll never treat technology the same after reading this book . . . and you shouldn't!"
—Joe Grand, electrical engineer and author of
Hardware Hacking: Have Fun While Voiding Your Warranty

"*When Gadgets Betray Us* opens your eyes to the implications of dependence on devices that don't always behave."
—Jeff Moss, Founder of Black Hat and DEFCON

WHEN GADGETS BETRAY US

THE Dark Side OF OUR Infatuation WITH New Technologies

ROBERT VAMOSI

BASIC BOOKS

A Member of the Perseus Books Group
New York

Copyright © 2011 by Robert Vamosi

Hardcover first published in 2011 by Basic Books,
A Member of the Perseus Books Group
Paperback first published in 2013 by Basic Books

All rights reserved. Printed in the United States of America. No part of this book
may be reproduced in any manner whatsoever without written permission
except in the case of brief quotations embodied in critical articles and reviews.
For information, address Basic Books, 250 West 57th Street, 15th Floor,
New York, NY 10107.

Books published by Basic Books are available at special discounts for bulk
purchases in the United States by corporations, institutions, and other
organizations. For more information, please contact the Special Markets
Department at the Perseus Books Group, 2300 Chestnut Street, Suite 200,
Philadelphia, PA 19103, or call (800) 810-4145, ext. 5000, or e-mail
special.markets@perseusbooks.com.

Designed by Brent Wilcox

The Library of Congress has catalogued the hardcover edition as follows:
Vamosi, Robert.
    When gadgets betray us : the dark side of our infatuation with new
technologies / Robert Vamosi.
        p.   cm.
    Includes bibliographical references and index.
    ISBN 978-0-465-01958-8 (hardback)
    1. Pocket computers—Social aspects.   2. Pocket computers—Health aspects.
3. Pocket computers—Security measures.   4. Software failures—Risk
assessment.   5. Computer crimes. I. Title.
QA76.9.C66.V36 2011
004.167—dc22

                                                                    2010043829

ISBN 978-0-465-03138-2 (paperback)
ISBN 978-0-465-02339-4 (e-book)

10 9 8 7 6 5 4 3 2 1

R060028 66783

CHICAGO PUBLIC LIBRARY

*For Laurie, my wife and fellow author, without
whom there could be no book*

CHICAGO PUBLIC LIBRARY

CONTENTS

INTRODUCTION

Why Gadgets Betray Us

In the seconds before the Pembroke-Swansea special came barreling down the railroad tracks to crush her car, Paula Ceely sensed something was wrong. Shortly after nightfall, the twenty-year-old college student had gotten out of her car in the pouring rain to open a gate blocking the road ahead. Ceely had used a borrowed TomTom mobile GPS unit to navigate the nearly 150 miles of rural road from Redditch, Worcestershire, in England, to her boyfriend's parents' house in Carmarthenshire, in Wales. It was her first visit. Judging by the illuminated GPS display on the dashboard gadget, Ceely was just a few miles shy of her final destination, and the road ahead should have been clear. When Ceely started opening what she thought was a farmer's access gate, common in rural England, she did not realize there were railroad tracks underfoot until the train, blowing its whistle, slammed into the tiny Renault Clio behind her. "I could feel the air just pass me," Ceely told the BBC shortly afterward, "and then my car just did a 360 degree turn on the tracks and was knocked to the other side."[1]

Ceely is not alone. In late 2006 and early 2007, a miniepidemic of mobile-GPS-related mishap stories was making headlines worldwide: A forty-three-year-old man in Bremen, Germany, turned left when instructed and drove his Audi right onto a tramway[2]; another twenty-year-old woman in England followed her dashboard GPS and drove her

Mercedes SL500 down a closed road outside the village of Sheepy Magna and into the swollen nearby river Sence,[3] and a man in Australia turned off a highway prematurely, driving through a construction site before stopping his SUV on the concrete steps of a new building.[4]

Reading these accounts one might conclude that consumer-grade dashboard GPS systems are, collectively, at fault. They're not.[5] Different vendors have sold millions of GPS-enabled gadgets for use in private airplanes, cars, and boats since the mid-1990s. ABI Research predicts that over 900 million people will use GPS navigation programs, available in both dashboard gadgets and via mobile phone, by 2013.[6]

Something else was happening when these commercially available GPS-enabled gadgets started hitting the larger population—something more fundamental. Instead of lifting our heads, looking around, and thinking for ourselves, some of us no longer saw the world as human beings have for thousands of years and simply accepted whatever our gadgets showed us.

Our need to know where we are is primal, and mobile gadgets give us that means in a way never before possible in human history. For many of us, myself included, it is an understatement to say that people today can't live without their technology. It's addictive. But in order to reach the masses, technology vendors have taken shortcuts. Software wizards whisk us through otherwise complex configuration settings, interfaces today have fewer and fewer options for advanced settings, and consumer goods are produced to be magic boxes whose internal components don't involve the end user. Along the way, we've introduced some unintended consequences.

What if our dashboard GPS gadgets deliberately misled us? GPS gadgets in our cars don't just provide navigation; they also warn us of upcoming road closures or accidents. What if they lied?

In the spring of 2007, Andrea Barisani and Daniele Bianco showed a video at the 2007 CanSecWest security conference in Vancouver, British Columbia, in which Barisani's 2006 Honda Civic GPS displayed a text alert warning of a terrorist threat near his home in Trieste, Italy.[7] This alert information doesn't come from satellites locked in geosynchronous orbit; rather, traffic alerts are sent locally via a ten-year-old radio protocol that satellite radio stations use to populate song names and details on dashboard entertainment screens. It didn't take long before someone figured out how to manipulate this protocol. The researchers' experiment was performed with a very limited scope so as not to interfere with other vehicles on nearby roads. And not all of the project was quite so hair-raising or serious. For their first attempt at injecting rogue messages into consumer GPS gadgets, the two Italian researchers popped up innocuous notifications such as "Bullfights Ahead."

Since roadside GPS alerts are not encrypted, anyone with the right equipment and knowledge of the signal used by the dashboard gadget could do this. The reverse is also true: Someone could block an emergency message in what is known as a denial-of-service attack. Thus, anyone with a low-power radio transmitter who knows the frequency used by a GPS unit can broadcast information—true or false—to passing travelers. While such ad hoc broadcasting is illegal in the United States, this is not the case in other countries.

Newer GPS gadgets use satellite-based alerts, which are much harder to spoof, although they also use unencrypted satellite signals. But older GPS units still relying on FM signals remain vulnerable to such an attack. Given that today we have a tendency to abdicate our common sense and simply trust these tiny wafers of silicon, if this book accomplishes only one goal, I hope it is that you will become much more skeptical about all the new gee-whiz gadgets coming our way.

1.

Not only can people send false information to our gadgets, they can also obtain personal data from us without our knowledge. The iPhone, for example, does not use GPS for its location services. Apple decided that tracing a phone's Wi-Fi Internet connection to a physical location holds significant promise over GPS.[8] Microsoft and Google have their own Wi-Fi location services. However, Wi-Fi is not necessarily superior to GPS for geolocation; it's just more convenient.

In 2008, a team of researchers in Zurich, Switzerland, found ways in which the Apple Wi-Fi location network could be compromised.[9] The iPad, iPhone, and iPod Touch gadgets query the nearest wireless access points—say an Internet café, a business, or a local residence—and transmit that information to a database, where it is correlated with a physical address (longitude and latitude). The Swiss researchers, however, fed this service incorrect information, telling the Apple service that the iPhone was in New York City when it was still in Zurich. But what if this vulnerability could be used with a more ominous intent?

Two years earlier, security researcher Terry Stenvold published similar findings in *2600*, a popular and well-known hacker magazine.[10] Stenvold found that he could steal someone else's smart device specifications— for example, the unique ID of a mobile phone or the unique hardware ID of a laptop—then upload that information to a location service and have the service tell him that person's current location.[11] Here technology could be used surreptitiously to track, for example, an ex-girlfriend's current location.

Already third parties can capture our location information and store it for an indefinite period. Have we considered the long-term consequences of this? How might a random trip to a seedy part of town look ten years later? What if it wasn't random? With enough data, what hidden patterns of obsessive behavior might emerge? Or, what if we could spoof our current location to make it appear that we are always at work when we are really not? Should we trust such location data? If this book accomplishes a second goal, I hope it will be to create an awareness of the various ways common gadgets can leak personal information.

2.

When Gadgets Betray Us, if you haven't already guessed, is a book about breaking things and not necessarily putting them back together. It is about hardware hacking, a relatively new area of research and concern: how our cars are vulnerable to attack, how our mobile phone conversations can be intercepted, how our contactless credit cards, driver's licenses, and passports can all be copied at a distance. The addition of basic authentication and strong encryption to most smart devices would significantly reduce the vulnerabilities described in this book;

yet, hardware manufacturers have so far shown little interest in securing their gadgets. Only by being more aware of the risk can consumers choose wisely.

There is a dark side, a secret life, to our smart phones, MP3 players, digital cameras, and new wireless laptops that most of us never glimpse—that is, until something goes terribly awry. We no longer read the manual before powering on; we demand intuitive interfaces that appear up and running right away, while often masking important security settings. Studies show consumers want complexity, perceiving gadgets with more capabilities as having more value, even if we don't understand how they work.

But how we use our gadgets is only half of the problem; the other half is the hardware itself. We fail to recognize that these same gadgets can fail. Or that they can be made to lie. Or track our every move.

"I don't trust many hardware devices. It's scary," said Joe Grand, president of the San Francisco–based Grand Idea Studio and former cohost of the Discovery Channel's *Prototype This*. "People using products today don't often think about what the gadget is actually doing. The product is helping you do whatever it is you want, but it might also be watching you or doing something nefarious."[12]

As a young hacker and now a hardware designer, Grand has seen both sides of the problem. For example, our desktop computers didn't leak personal data until they were connected to the Internet. "Once you allow a system to connect to a network," Grand said, "then it's open to a whole different side of attacks. You don't even need to be a hardware hacker. You could be a network hacker or a software hacker. Now you're in a brand-new world."

The convenience of having little chips inside a toaster to make a perfectly browned piece of bread is nice, but when we add to the toaster the ability to connect to the Internet and order more bread, that opens us up to unintended consequences. Your toaster could someday become the victim of a denial-of-service attack, unable to operate because some remote party has reprogrammed its firmware. This might sound funny, but when applied to other gadgets such as implanted medical devices, it isn't.

Grand cited another example: "Monitoring your driving activity could be used by automobile manufacturers to improve the car's performance and safety, but it could also be used by insurance companies to verify you weren't speeding when you crashed into that tree." Grand isn't being paranoid; he's right. "I think your microwave probably isn't tracking how often you use it," he said, "but your car might be."

The hardware industry's rebuttal that the exploits depicted in this book are beyond the average criminal's resources is no longer true. The more complex technology becomes, the easier a gadget is to break. Cybercriminals don't necessarily have to know more than we do about a given technology; they just need to know how to defeat it. A group of young carjackers in Indonesia, for instance, will, out of frustration when confronted with a state-of-the-art biometric-protected luxury auto, simply cut off the victim's index finger and use the severed digit's fingerprint to steal the vehicle. In another take on this criminal realm, a streetwise thug in Prague who today uses a laptop with software downloaded from the Internet to steal cars is essentially no smarter than the thief who used a screwdriver and a pair of scissors to hot-wire a car ten years ago.

Thanks to a combination of Moore's Law (which says the number of transistors placed on a chip will double every two years) and the passage of time, the costs of smart device attacks have come down dramatically.[13] For example, an individual with a modern dual-core processor in a Dell laptop, loaded with the right software, can defeat a twenty-year-old encryption algorithm not in a matter of days but in a matter of minutes.

Another hardware industry response maintains that public vulnerability disclosures from security researchers help criminals. But whether or not they are disclosed, the smart device flaws are probably already known by the criminal community. We speak of exploited vulnerabilities previously unknown to computer software vendors—"zero days"—being traded by criminals openly on Internet black markets; without a good-guy security research community finding and reporting these today, the same is happening with smart devices. However, smart device manufacturers have sometimes threatened researchers mentioned in this book with legal action if their work is disclosed. To protect us all, that culture needs to be changed.

Finally, the hardware industry cites the considerable resources needed to defeat their products. Except attackers today do not need sophisticated computer coding skills; gibberish code can also make a gadget fail, and in a pacemaker that can be lethal.

Security researcher Deviant Ollam, a lock picker and physical security consultant, echoes the concerns of others in saying smart device manufacturers are where software vendors were with security concerns twenty years ago. The lock manufacturers tend to belittle researchers like him, he said, "which is fine if you are making an unpickable product. But no one is making an unpickable product. Everyone has a weakness. And you should be embracing people who will tell you about it."[14]

There is a fine line between useful and malicious hacking, which blurs the distinction between "good" and "bad" in this book. The idea that someone like Joe Grand or Deviant Ollam can make an honest living breaking the gadgets that you and I use every day might seem ludicrous; yet, there is a desperate need for this research and for the security conferences that openly discuss the varied activities of cybercriminals. If this book accomplishes a third thing, I hope it will be to start a constructive dialogue between the manufacturers and the research community around security.

3.

In the physical world, we're quite adept at sensing danger. Our ears prick up at strange sounds; our skin tingles when something doesn't feel quite right; we notice subtle body language in a stranger that makes us suspicious. We are hardwired to recognize the authenticity of another human being by a look in the eye or a firm handshake; yet, most of our authentication today occurs digitally, by voice, text, or e-mail, via subatomic particles moving at the speed of light. We don't know when someone tries to extract personal information from us or eavesdrop using our mobile phone. And whenever we do use technology to authenticate another person, too often we invest in simplistic filters or imperfect biometrics that result in many false positives. In the most extreme cases, these flawed biometric systems even send innocent people to jail for crimes they did not commit.

With technology we simply haven't evolved our survival instincts. We often make leaps of faith with new technologies based on very few criteria. Gadgets today are so complex that often we're just happy to get a new product working—and too intimidated to change its default settings afterward. Yet, we *should* change those settings.

Gadget manufacturers that simplify their complex technologies only give us the illusion of control, and this in turn opens the door to greater risk. Remember Paula Ceely? We may not always be aware of this danger, placing our confidence in electronic gadgets that greatly simplify the real world or entrusting them with our fates—or with personal information that we once locked away in safety deposit boxes. Yet others will use our naiveté against us.

With gadgets we believe that a new technology—such as antitheft circuitry in our cars—somehow trumps all the real-world experience we've gained over the years. Instead, we should be layering our defenses— such as parking in well-lit spaces, using a physical lock on the steering wheel or brake pedal, and applying antitheft technology—*adding* rather than *subtracting* security. But human nature is such that we prefer convenience over effort. We lock only the outermost doors on our houses because 90 percent of the threat exists there. We may have sensors that tell us whether our windows have been opened, but they won't tell us whether they have been broken out to let a criminal in. Similarly, we entrust the security of our cars—the most expensive purchase we make next to buying a house—to a single beep-beep. With keyless entry and remote-ignition cars, physical keys have morphed into a single gadget that both unlocks and starts the car with a touch of a button. But does this one gadget make the car any safer from theft?

While we may patch software flaws in our computers, we are not yet accustomed to patching security software flaws within our TVs and DVD players. The mobile phones we use are increasingly asked to provide more services—fast Internet connections, text messaging, applications, e-mail, Web browsing. Yet, by recycling older networking protocols and standards that have been attacked in the past in order to meet future demands, we're left vulnerable. Are we patching our mobile phones to new vulnerabilities? Are we even thinking of our mobile phones as computers?

And it is not just attacks we should be aware of. As a result of our misplaced trust in our gadgets, we're leaving behind a trail of electronic bread crumbs that, when viewed in the aggregate, may suggest patterns others can exploit. Photocopiers remember our sensitive documents, and photos posted to the Internet reveal our location at the moment they were taken, invading our privacy without our knowledge. The consequences of having a tollbooth transponder monitor our daily comings and goings escapes most of us—until a divorce lawyer uses that rather bland data to construct a rich narrative about how we were, on certain afternoons between 4 and 6 p.m., having an affair.

By adding contactless broadcast systems to our worker-access badges, driver's licenses, and passports, we're speeding up the authentication process—but we're also creating new kinds of identity theft. Cloning wireless signals is easy. With no authentication and often with little or no encryption, or with trivial encryption, I can become you without ever coming into physical contact with you or your papers or effects. Additionally, retailers are embedding RFID tags in the products we buy. While no personal information is revealed and the tags themselves have only serial numbers, collectively these product tags create a unique electronic proxy that becomes a de facto consumer, and this can now be tracked from store to store.

These electronic proxies, sometimes taken without our knowledge, are, however, no more accurate as representations than our biometric information, sometimes knowingly given. A fingerprint is often thought to provide a one-to-one match—except fingerprint matching is a myth.[15] Biometrics miss much of the data that makes us unique, just as GPS gadgets cannot possibly account for every real-world roadside detail.

Perhaps one way to contend with all of this technological change is to converge all our disparate electronics down to just one gadget—

if, on this one gadget, we also build in security first rather than bolting it on later. We seem never to forget our mobile phones; yet, we sometimes forget our car keys and our wallets. Perhaps such mobile phones, if designed securely, could hold our personal contacts, our to do lists, our credit cards, our music, our photos, and more. Today smart phones are such a mobile gadget. And unlike a physical wallet, if ever lost or stolen, a smart phone can, with additional software, be remotely locked or erased, thus rendered useless to a thief. But before we can successfully integrate technology fully into our everyday lives and navigate the new "Internet of Things," we'll need to change our behavior around gadgets and, in general, become much better informed about how they work and their various vulnerabilities.

In the seven chapters that follow, we'll travel from the streets of Berlin, to Prague, Johannesburg, Los Angeles, New York, and elsewhere. We will meet people who have experienced firsthand how gadgets can and have betrayed us. The point is not to scare everyone away from technology—we're too committed already—but to promote its intelligent use in our everyday lives and to effect wise behavior in order to minimize personal risk.

CHAPTER ONE

A False Sense
of Security

For most of us, that familiar beep-beep as we walk away through a parking lot or garage is enough assurance that our car is both locked and safe. Often a tiny flashing light on the dashboard also alerts would-be criminals that the car is protected by the latest form of antitheft security. And for the most part, that is true; a sophisticated set of encryption and electronics is at work inside. However, don't be surprised to find your state-of-the-art, antitheft-protected vehicle stolen. Complex technology doesn't necessarily raise the barrier for entry for cyber-criminals; sometimes it does the exact opposite.

Just ask Czech-born Radko Soucek, a streetwise career car thief and unlikely example of a high-tech criminal. Soucek, now in his thirties, has been stealing cars since age eleven in a country that holds the un-enviable reputation of having ten times more car thefts per year than any other European nation, according to the International Association of Auto Theft Investigators.[1] Czech officials attribute most of the 51,000 thefts per year in that country to thieves who work in teams stealing cars, forging registrations, and stripping parts—organized crime, by any other name. Soucek works by himself. "You leave your car, lock it and walk around it toward your house: That's how long I would need to take it," he told the *Prague Post*.[2]

In the 1990s, as more and more European automotive manufacturers started incorporating computer technology into expensive Mercedes, BMWs, Ferraris, and Porsches, Soucek realized he could defeat the manufacturer's antitheft software with his own. Lacking any formal computer training, he uses Internet-provided software, which is rapidly becoming available in Prague and elsewhere. From within Prague-Ruzyně Prison, Soucek said that twenty years ago all he needed was a pair of scissors to steal any Italian sports car. "Now you need a lot more technology." He said he no longer uses dime-store implements; today he uses a laptop.

Gangs, like those operating on the streets of other European countries, often search for and steal particular high-end makes and models of cars. By specializing, it's possible for these gangs to guess, through sheer trial and error, the electronic antitheft codes found in keyless entry fobs. Another possibility, one that's more likely, is that they already know the vendor's proprietary code algorithm (it was either stolen, purchased, or provided by an insider or someone within a dealership).

Codes used by these antitheft systems do not make us more secure; they make us complacent. We trust in them so much that we forget commonsense lessons, such as parking in a well-lighted spot, hiding valuables, or using an auxiliary locking mechanism on the wheel or the brake. We assume the high-tech solution is somehow better than past experience. We have become careless with our cars and our sense of what's secure.

Security works best in layers. As we will see, antitheft technology in cars is actually going backward; instead of adding security, manufacturers are decreasing it by providing greater convenience to the driver. And we're also to blame. We're so confident in our belief that high technology is better than common sense that we'll ignore the condition of the neighborhoods we park in and do without the benefit of The Club or other steering wheel–locking gadget.

Yet, the auto insurance industry disagrees. Clearly something has resulted in a decrease in auto thefts in the United States in the last few years. Preliminary U.S. Department of Justice figures for 2009 show a remarkable decrease of 17.9 percent in auto thefts.[3] This follows a 12.7 percent decrease in 2008, 8.1 percent in 2007, 3.5 percent in 2006, 0.2 percent in 2005, and 1.9 percent in 2004. The National Insurance Crime

Bureau (NICB), a nonprofit organization that follows car theft, finds a similar six-year decline. NICB data show that 83 percent of the 366 metropolitan statistical areas within the United States reported lower thefts in 2009 than in 2008.[4] Much of this decrease, I think, is the result of education and legislation and not, as the insurance industry claims, the increased use of antitheft gadgets.

Although it is impossible to say exactly how many auto thefts are the direct result of laptops emulating the digital codes issued by a standard key fob, it is more than a few.[5] When he was arrested in 2006, Soucek had the data for 150 stolen cars on his laptop. "You could delete all the data from your laptop, but that's not good for you because the more data you have, the bigger your possibilities," he said.[6]

So, how hard is it to use a laptop to steal a new car?

First, we need to understand what's happening when we unlock the door, insert the metal key into the ignition, and start a car today.[7] Most cars use a keyless remote entry fob: You push a button, and the resulting radio signal either locks or unlocks the car's doors; in some models, it opens the hatch or trunk. Using a tiny battery, the fob can broadcast a coded signal up to one hundred feet in order to make contact with the car, generating the beep-beep and the flash of headlights that audibly and visually identify your car in a crowded parking lot. The fob and car wirelessly exchange a series of nanosecond challenges and responses. If the car receives the expected code, it performs the function.

For added security, these codes are rolling, or what the industry calls hopping, codes. Both the keyless fob and the car use the same pseudo-random-number generator following a proprietary algorithm. When you lock or open your car door, both the car and fob store into memory the next code. If you hit your key fob while away from your car, the car and fob will fall out of sync. The car receiver solves this by accepting any of the next 256 possible codes. If you press the fob 257 times while far away from the car, however, you may not be able to resynch the fob with your car. It's important to note that the key fob in this case only controls entry to the vehicle.

Once you are inside, a second antitheft technology, a static vehicle immobilizer chip embedded within the plastic base of the key, becomes important. Immobilizers in the United States have been cited for the sharp decrease in auto thefts in recent years. Unlike with keyless entry, the immobilizer's radio frequency identification (RFID) chip must be queried, or "energized," externally by the car. After you insert the immobilizer key into to the ignition block, a transponder within the car (usually near the steering column) energizes and queries the chip inside the metal key. In exchange, the energized immobilizer chip broadcasts a low-frequency code. The broadcast distance between the key and the immobilizer is only a few inches, so the key must be in the ignition slot.

Once the chip inside the physical key is validated, the immobilizer system unlocks the rest of the electronic systems in the car. Older cars use what are called fixed keys (one code per vehicle), while cars made today randomly generate and store new immobilizer codes after each use. Today immobilizer systems are no longer separate components of the car but bundled within the electronic subsystems.

Even without validation of the immobilizer chip, a car can be driven a short distance before locking up. A valet key, often provided by the dealer as a third key, lacks an immobilizer chip. The valet key exists to allow the valet to park the car a short distance away, not drive off on the freeway.

These two technologies—keyless entry and vehicle immobilizer chips—form the basis of most high-tech antitheft technologies in cars sold today. Both rely on RFID codes exchanged over the air. The flaw, if any, is that most cars use only forty-bit encryption for this; upon introduction in the 1990s, this was sufficient, but it is no longer adequate. The more bits of encryption, the harder it is for someone to guess or break the code. The more bits, the more processing time and resources you'll need to do so. Forty bits used to take days to crack; now it takes much less time. Today 256 bits is considered strong encryption, but it is doubtful you'll find a car on the street with that level of crypto.

That's because chips today are much faster. Back in 1965, Intel's Gordon Moore famously wrote in *Electronics Magazine* that the number of transistors on a given chip would double every two years.[7] This exponential growth in computer processing power has lead to the more powerful, yet less expensive, computers we have today. And because of trivial flaws inherent in the basic underlying design of some digital signature transponder (DST) devices or key fobs—be they manufacturer or third party provided—some cars today are more susceptible to laptop car thefts than others. And as Soucek demonstrates, it doesn't take a genius to realize that.

In Radko Soucek's favorite film, *Gone in Sixty Seconds*, Nicolas Cage's character has a holy grail—one car that he has longed for but never stolen.[9] For Soucek that car is the Mercedes Maybach. At prices close to US$500,000 each, only a few exist within the Czech Republic. Soucek knows where those cars can be found.

He has no doubt he could defeat the antitheft system in the Maybach if he got the chance. But Soucek has a much more immediate concern. In 2006, he was convicted of stealing at least 150 cars in Prague over a six-month period. He is currently serving a prison sentence.

That Soucek was so prolific and so successful is no surprise, given that he kept all his previous antitheft code keys on the hard drive of his laptop. He had to. As he successfully crunched the numbers of one car, he used the previous successful sequence to calculate the code of the next car from the same manufacturer. This is smart for a career criminal since Soucek, working independently of a gang, could build his own database, then learn to anticipate each new car's individual code sequence through trial and error. But it's also very risky. When authorities finally nabbed Soucek, they found more than enough evidence on his laptop's hard drive to arrest him.

When Soucek gets out, he's already got plans. He told the *Prague Post* he's outgrown the Czech Republic; he might just take his act to France or Spain.

Or to the United States.

He said, "I would like to take my activities abroad to show them a little of how it's done."

1.

Shortly after purchasing their new 2003 Honda Civic Hybrid, Brad Stone and his wife woke to find their new car (affectionately known as

Honky) missing from the San Francisco street where they'd parked it the night before.[10] Stone still had his original set of keys. So did his wife. They even had their third valet key. Yet, someone had stolen their supposedly theft-proof car.

The Stones' Honda reappeared a few weeks later not far from where it had been stolen. In fact the couple had called the police to inquire about its status only to learn that their car was accruing parking tickets a few streets away.[11] Honky had not been stripped; nor were there obvious signs of damage to the dashboard or engine. Yet, the car was out of gas, its interior littered with cigarette butts—someone had obviously gone joyriding. Stone, now a senior writer for *Bloomberg Businessweek*, decided to investigate. Over the years since, he has written several articles for various publications detailing how someone might steal an antitheft-protected car.

There are two ways to go about stealing a car. One is destructive, or "smash and grab"—throw a brick through the passenger window, cut some wires in the dashboard, and hot-wire the car. The other is subtler, more surreptitious; there is no visible damage, sometimes not even evidence a crime has been committed, leading some insurance companies to allege that the owner staged the event. The Stones' incident was a surreptitious one, often the mark of a professional car thief.

Critics faulted Stone's auto-theft reporting as neglecting to mention several physical controls built into all cars. These include an ignition lock, a steering column lock, and a protected starter switch for the ignition. According to North American Technical & Forensic Services's Robert Mangine, "You cannot physically defeat the steering column without leaving damage to the components or detectible [*sic*] marks or scars to the ignition lock."[12] Mangine argues that existing physical security built into the ignition block would appear to add a valid second layer of security to any auto, whether or not it has an electronic transponder antitheft system. "So," Mangine concludes, "if the transponder system is compromised, by-passed, or re-programmed, you would only have succeeded in defeating one-half of the vehicle protection."

A second critique by Jeffrey Lange simply dismisses Stone's reporting due to lack of peer review, finding that "any and all opinions cited by Mr. Stone are considered without evidentiary or scientific value and, as such, none have 'forensic' value."[13] Lange states that "although there is some accuracy to the information reported, the position of the author is based solely upon specific data from questionable sources regarding a limited number of vehicles." He concludes by adding, "To suggest that all transponder immobilizer systems can be circumvented as part of a vehicle theft is inaccurate and irresponsible."

Clearly, Stone's reporting touched a nerve within the insurance industry. In a *Wired* article, Stone cited a theft of a 2003 Lincoln Navigator from a Target store in Orange, California.[14] The owner, Emad Wassef, reported the theft to the police and his insurance company. But when the vehicle turned up near the Mexican border "minus its stereo, airbags, DVD player, and door panels," the insurance company balked at paying for the loss. Their forensic investigator found that the ignition lock had been forcibly damaged, as had the steering column lock, but the transponder system was intact. Since Wassef, like Stone, still had his original keys, the investigator concluded that the loss did not happen as reported. In other words, Wassef himself may have driven the car to the border and used tools to damage it, a charge he denied.

So, without a set of keys, how could a criminal defeat physical locks and electronic transponder systems without damage and still drive the car? Stone found that several car manufacturers provide a few different ways to disengage the vehicle immobilizer and start your car in the event that you lose all your original keys. Honda includes such a fail-safe method for starting Stone's Civic, one that involves a particular sequence of presses and pulls on the emergency brake. This odd, shamanistic dance in the driver's seat will allow you to drive as long as needed. Fail-safes exist with other makes of cars as well. In theory, these fail-safe systems were designed get your car back to a dealership, where you could order a replacement set of keys. In practice you could live without the metal immobilizer key, although you'd look silly in the parking lot. Unfortunately, this manual process is well-known within auto-theft circles. To learn the specific press-and-pull sequence for your car, a thief needs only to obtain its vehicle identification number (VIN), which is visible from the outside of the car.

With the VIN number, thieves can use locksmithing resources to cut a shiny new metal key for that specific automobile. Using this non-immobilizer chip key is the same as having the valet key; it only allows thieves to open the door and drive the car a short distance (say, into the back of a semi truck), at which point they can further compromise the immobilizer system or begin stripping the car at a remote location. But Stone's ignition block and steering column lock had not been damaged. How could that be?

Say you want to sell your car and can only find one of the original keys. Or what if you simply lose all your keys? It turns out that ordering a replacement set is possible, but it will cost you.[15] A metal key may cost $12 to replace, while a "smart" chip–enabled key costs $150; some cost as much as $335 per key. If all the keys are lost, then the computer within the car will also have to be replaced, costing even more.

But thieves aren't going to pay for new keys or computer systems, and Stone's original keys still worked. His thieves had to use another method to start the car. Again, it's likely they went old-school.

Stone learned that some car thieves still use what are called jiggle keys to bypass the keyless entry fob—sawed down keys that will trip most metal tumblers and manually open the car door. This allows the thieves inside to root through the glove compartment, where many people leave their owner's manual and the valet key.

Whether thieves use a VIN to create their own key or rely upon manual pushes and pulls really isn't the point. Antitheft-protected cars have been and will continue to be stolen. Perhaps mechanical and electronic locks are not all that secure.

2.

In his classic book *Hackers* (1984), Steven Levy made the connection between computer hacking and lock picking. In order to get after-hours access to the equipment they needed, according to Levy, computer geniuses at the Massachusetts Institute of Technology in the

1960s found ways to defeat the locks within the various engineering buildings. These were surreptitious entries; the hackers only wanted to borrow the tools and return them unnoticed. As Levy described it, younger members of the early hacking clubs would be tasked with creating master keys for the floor where others needed access. It got so bad that the head of security later said he opted for a policy whereby the hackers could do what they wanted as long as they didn't talk about what they were doing.[16]

Lock picking, like computer hacking, is both a science and an art. Metal locks, like computer code, present enthusiasts with a real-world puzzle. Locks use a series of pins or tumblers that, when pressed in the correct sequence, allow the mechanism to open. While there's a little bit of science—for instance, you have to know how a given lock is constructed—a great deal more chutzpah is required to actually pull it off. Think of a safecracker in an old movie with his delicate fingers "listening" to the hardware before him. Some people have the touch. Marc Weber Tobias and Deviant Ollam are two security researchers who have the touch.

Conventional pin tumbler locks are constructed using what's called a keyway and a series of different lengths of pins set along that path. A key is cut to match the sequence of pins so that the pins are all at the same level, allowing a cylinder containing the keyway to turn. For example, a peak on a key corresponds to a recessed pin, whereas a valley corresponds to a distended pin. The number of pins influences the number of overall combinations possible. For generations, key and pin combinations have been sufficient with little or no real variations.

From the mid-Atlantic region of the United States, Deviant Ollam is on the board of directors of The Open Organisation of Lockpickers, or Toool, an organization of enthusiasts and researchers that has become a fixture at security conferences like the annual DefCon security conference in Las Vegas, Nevada.[17] Unlike Tobias, a lawyer from South Dakota who has been interested in the art of lock picking since the age of fifteen,[18] Deviant is not part of a locksmith guild; he is what practicing locksmiths might refer to as a sport picker, although he is a professional security researcher.

Deviant describes locks as a "way to provide (in theory) rapidly-deployed and easily-removed barricades that alternately restrict or allow easy passage or access to a sensitive resource."[19] Whether to prevent access to a car or a stash of jewels, a single lock by itself is not enough.

This introduces two truths about locks.

One: No security system can deter attackers all the time—you can only throw enough obstacles in their path so that they'll either give up or make so much noise or take so much time that they'll be caught. Thus, relying on any one method of security is a recipe for disaster.

Two: All hardware has flaws, even locks. Of course, unlike with software, however, a hardware flaw is almost impossible to patch with an upgrade. Thus, a flaw in hardware is usually permanent unless the whole device is replaced.

To defeat locks, Tobias employs a few simple tricks. One is to obtain a master key, one key that can open several locks within a facility. He has shown at security conferences that one need only trace or even photograph the key in order to fax or broadcast the design to others who can then use the image to make their own master key. Tobias and a colleague once demonstrated before a live audience how he could open a $100 Medeco-brand lock in a matter of minutes. While impressive on its own, the demonstration also conveyed an implicit threat: These hundred-dollar locks are used to safeguard embassies and even the White House.

After various media reported an assortment of secure facilities around the world were now vulnerable, Tobias argued that they always had been and that this publicity gives "good guys" a chance to catch up.[20] Deviant agrees: "If you put enough sunlight on anything, more honest people are going to find flaws than dishonest people," he said. "The honest people will tell you about them if you let them. The dishonest people will try and exploit them, but there are fewer of them."[21]

The lock industry does not see things that way. Although Tobias has repeated his exploit a number of times before a number of different audiences, Medeco has refused to acknowledge his feat publicly.[22]

Deep down, mechanical locks are often not as unique as the manufacturer claims, said Deviant Ollam. Indeed, after hundreds of years, how unique is a metal lock? Even under the best of circumstances, any lock will have some mechanical flaw. Deviant Ollam, Tobias, and others are simply tweaking those preexisting flaws. That's at the heart of surreptitious entry and lock picking: exploiting existing flaws. And some are ridiculously easy. How many of you know that an ordinary Bic ballpoint pen was enough to defeat the popular Kryptonite brand U-locks used on bicycles?[23]

Back in September 2004, Chris Brennan, a twenty-five-year-old San Francisco bike rider and network security consultant, wrote a post to a site called Bike Forums expressing outrage at the rise in bicycle thefts in San Francisco.[24] In it Brennan explained the process he'd used to defeat his own brand-new Kryptonite Evolution 2000 lock.[25] At the time, I was commuting to work in downtown San Francisco and trusted my handy Kryptonite lock implicitly. After reading about Brennan's revelation in a local paper, I, too, was able to use a random Bic pen to open my Kryptonite lock.

The problem, as Tobias explained later, was that the Kryptonite lock used an axial pin tumbler that happened to be the size of the average Bic pen barrel. Furthermore, the soft plastic on the pen tip molded to the inside of the lock, making it easy for anyone to pop a Kryptonite Evolution 2000 lock open. Deviant has pointed out in lectures and training sessions that the design of this particular mechanism was replete with engineering faults from a security perspective, such as weak springs and pin chambers so deep that the entire array of tumblers could be pushed almost completely out of the plug. Tobias had first heard of the Bic pen unlocking trick at a computer security conference earlier that summer, but it wasn't until Brennan's post published the details of the hack that the vulnerability went public.[26]

It wasn't just Kryptonite locks that were vulnerable. The basic axial pin tumbler (or tubular lock) had been used by a variety of different lock manufacturers for nearly half a century.[27] A 1992 United Kingdom publication, for example, reported the axial pin tumbler lock flaw, and apparently the system was known to be vulnerable within the lock-picking community before that.[28]

Given that these flaws are known, what's keeping the lock industry from taking them seriously? Tobias said that the vulnerabilities he's discovered and disclosed are the result of "security wars" between manufacturers, criminals, hacker locksmiths, and others. He states that Medeco "claims that its locks meet or exceed all applicable high security standards," but he adds, "What if the locks can easily be opened by methods not contemplated within the standards?"[29]

On his blog Tobias wrote, "Lock manufacturers are very proficient at making locks work properly. That is what we refer to as mechanical engineering. Unfortunately, the engineering groups for some of the world's most respected companies may not, in our opinion, have the requisite skills when it comes to security engineering. . . . In other words, sometimes they cannot figure out how to open their own locks without the correct key."[30]

Tobias said that some lock vendors simply don't believe their gadgets were designed to be picked, attacked, and otherwise tampered with. That's remarkable since the presence of a lock implies there's something valuable within. Tobias, as a lawyer, has posed the question of liability to the general legal counsel of several lock manufacturers, who respond in turn that their respective companies do indeed guarantee the locks against "normal use."

What is normal use? As I discuss in the following chapters, the manufacturers, software vendors, and companies that produce the gadgets we use in our everyday lives often do not realize or imagine the many unintended ways in which people actually do use their products. Like Paula Ceely, we as consumers of electronic gadgets have high expectations for what a given piece of technology ought to do as opposed to what it was designed to do—which may again be different from what it actually does do in practice. Sometimes the design is flawed from the outset, built onto a system found to be insecure.

To remedy this, Tobias proposes that lock manufacturers hire competent security engineers, people who know how to break things in order to implement a secure design. He would also like to see these locks tested against a variety of different security threats. At the moment, top lock manufacturers use Underwriters Laboratory and Builders Hardware Manufacturers Association and the American Society for Testing and Materials to test their locks. These are not security organizations, however; they're manufacturing organizations.

Tobias makes a case that lock manufacturers should be held legally liable for defects in their products that they know about or become aware of during the lifetime of the product. That's a radical claim. Unlike software, for which updates can be pushed down to the user at any time via the Internet, hardware in the physical world is much more expensive to replace. First, a replacement unit has to be designed, then actually manufactured; then it must be shipped to the public, which must install it—often at a cost. With autos, safety upgrades are performed via vehicle recalls, with a trained mechanic installing the part. With smaller gadgets, that is not always possible.

Tobias would like to see lock manufacturers—the high-end ones at least, those who market their devices as "high security" and sell such locks for more than a few hundred dollars each—replace defective designs. In response, Clyde Roberson, Medeco's technical director, told *Slate*, "When you buy a lock, you don't buy a subscription."[31] In other words, locks—even high-end locks—are sold as is.

This is an interesting problem. When the Kryptonite lock flaw surfaced a few years ago, the bicycle lock company not only fixed the problem going forward but also replaced a number of the defective locks retroactively.[32] It had to. Kryptonite, a division of Ingersoll-Rand, is enormously popular with the biking community.[33] The company's reputation depended on the trust of bicycle owners worldwide. The company soon switched to the more secure disc-style lock cylinders. But it took public disclosure to get them to make that switch.

After Tobias went public with his findings, however, locksmith organizations, such as the Associated Locksmiths of America Inc. (ALOA), tried to silence him. It was as though Tobias was a magician about to sell out his trade and reveal how to do some complicated trick. The criminal world already knew all the vulnerabilities that Tobias was making public.

By discussing these vulnerabilities, Tobias forced a serious deliberation about design flaws with an eye to fixing them. Yet, fellow locksmiths were not amused. In 2004 an ALOA member filed a formal grievance against Tobias, and in 2007 the organization as a whole threatened to expel him for presenting security weaknesses to people of "questionable character" (i.e., the hackers he addressed in appearances at DefCon). Tobias told *Wired* magazine, "They're pissed because I keep telling them that it's not a guild and that there are no secrets. It's called the Internet—duh!"[34]

Deviant Ollam agrees. "There are friends of mine, locksmiths—especially in Europe and the UK—who have been outright banned from organizations," he said. "They've had their memberships revoked, just like Marc. It's a common theme. Anyone seen as collaborating with the public, not even collaborating with criminals . . . that's as close to violating gospel as you can come in the locksmith world. It makes no sense to me or anyone in the tech security world. It just doesn't add up."

Unfortunately, as we will see later in this book, there are other examples in which the messenger (often a security researcher or academic) is threatened with legal action by the very industry that could most benefit from his or her work.

3.

In recent years auto manufacturers have removed the physical key from some cars entirely; in the process, they have removed the half of the security equation that Robert Mangine and others alleged that Brad Stone had not accounted for in his articles. Auto manufacturers and insurers may insist these new, purely keyless, electronic technologies are equal to (or better than) older mechanisms and entirely invincible to theft. Yet, world-famous soccer star David Beckham might tell you otherwise.

Beckham holds the distinction of having not one but two very high-tech keyless BMW X5 SUVs stolen off the streets of Madrid. One theft occurred in broad daylight.[35] Before he signed on with the Los Angeles Galaxy soccer team, Beckham had played for many years with Manchester United in the United Kingdom, then Real Madrid in Spain. In living around the world, Beckham has taken his cars with him, and his collection includes an Aston Martin Vanquish S, a Hummer H2, a Lincoln Navigator 4 x 4, a Range Rover, a Lamborghini Gallardo, a Bentley Arnage, a Rolls-Royce Phantom, and an Audi A8.[36] While living in Spain, Beckham bought an armor-plated BMW X5 SUV for his day-to-day driving. That may sound odd, but while they were living in England (in fact, perhaps one of his reasons for leaving that country), his wife, Victoria, and young sons became the target of a would-be kidnapping and extortion scheme.[37] Concerned for his family's safety, Beckham bought the family two armor-plated vehicles in Madrid.[38]

One was a Range Rover V8 Vogue with a bulletproof door.[39] The other was the BMW X5 SUV. As long as Beckham had the original key fob on his person, he needed only to lift the door handle to unlock the car. The car queried the key fob in his pocket to arm and disarm the security system, open the hatchback, make minute adjustments to the seat positions, or allow the driver (once inside) to press a button to start the car. It's called a keyless ignition system, but it includes entry as well. The convenience of operating a car without fumbling in your purse or pocket for keys is a godsend to any parent with small children in tow. Even for adults without kids, the ability to walk up, get inside, tap the brake, and simply touch a button to start the engine is very futuristic. But now we've gone from having two separate encrypted chips (the digital signal transponder and the ignition immobilizer) to just one. From a security perspective, that's not good.

Madrid, like Prague, is a European hotbed of car theft, with an average of fifty luxury cars stolen every day.[40] In November 2005 Beckham's first BMW X5 was stolen.[41] An associate of Beckham's had parked the vehicle in the La Moraleja region outside the Gran Madrid hotel not far from Beckham's villa. The associate later told authorities that he forgot to engage the extensive antitheft security system. No matter the high-tech solution, the security doesn't work if you forget to engage it. The car, missing for more than a year, reportedly showed up in Macedonia, along the border with Albania.[42] Authorities in Spain speculate that Beckham's X5, which seats seven, was stolen by professional car thieves, perhaps without their knowing to whom it belonged, then used for human smuggling in and out of various European countries.[43]

By then Beckham had obtained another X5.[44] One afternoon in the spring of 2006, while driving in suburban Madrid with his three boys, Beckham stopped at the Diversia Center at Alcobendas for lunch. Perhaps he thought that lightning would not strike twice or that no one would be bold enough to steal his car from a crowded shopping mall parking garage. Upon returning from lunch, however, the soccer star discovered that thieves had once again stolen his latest vehicle.[45] Beckham afterward decided to get an Audi.[46]

By removing one layer of security—a metal key for either entry or ignition—we've reduced our electronic auto antitheft technology to a single point of failure. Yet, without a physical key for entry or ignition, how were the thieves able to enter and drive away with Beckham's second car? There are two theories.

Keyless ignition systems are not new. Back in 1913, the U.S.-made Locomobile became one of the first cars to use push-button electronic starters. Locomobiles were produced by a company owned by John Brisben, editor and publisher of *Cosmopolitan* magazine. As Henry Ford's Model T (which favored keyed over push-button ignitions) began to outsell Locomobiles, the keyless ignition idea fell by the wayside. In the early 2000s, the idea was rediscovered as cars from Audi, BMW, Mercedes, Nissan, and Toyota began to offer keyless or push-button ignition systems as options.[47] Called Comfort Access by BMW and Start System by Toyota, the two modern keyless systems are basically the same: A key fob in your pocket or purse contains a transponder attuned to your particular vehicle.

Here again, human nature caught up with the idea as losing the original keyless ignition key fob meant taking the car into the shop. Unlike the vehicle immobilizer system, the keyless ignition system has no failsafe. Someone must have the key fob (or something that emulates it, like a laptop with some radio-enabled peripheral devices) in order to defeat the built-in security.

Some industry analysts, such as David Bell of SBD Ltd., a British automotive consulting agency, discount media stories about laptop-related auto thefts. Bell offers a far simpler solution: He said recent increases in theft among high-tech cars protected with antitheft technology can be attributed to fraud at the dealerships, with thieves obtaining original key fobs through illicit means. "Dealerships are not skilled in fraud prevention and only have limited responsibility for the consequences if the worst does happen," said Bell.[48]

He may be right. One car-theft ring operating in the San Francisco Bay Area during 2005 provides an illustration. Naheed Hamed, twenty-four, was arrested at his home outside Campbell, California, on charges of causing a hit-and-run accident and evading arrest after slamming his Mercedes into a tree at over one hundred miles per hour.[49] Eventually Hamed was also charged with grand theft auto after it was discovered the 2006 Mercedes he was in at the time of the accident had been reported stolen earlier in the day from a Mercedes dealership in Monterey, California. Hamed, it turns out, was out on bail at the time for stealing yet another Mercedes from another dealership two weeks before. He was fined $200 and given thirty-six months probation.[50]

How Hamed stole the two cars was clever, if low-tech. He would walk into a Mercedes dealership, express interest in a particular model, take it for a test drive, and then, when the salesperson went around back to get some paperwork, swap the keyless ignition fob with a blank one. Later, when the dealership had closed, he would return, activate the car with the original key fob, and drive off.

As we're about to see, however, I think a valid case can still be made for the laptop method of stealing keyless ignition cars. The main barrier, say most critics, is that you need to replicate the code sequence from the typical car key, which can have up to 1 trillion different combinations. So, how might you be able to generate those and still pick the right code? With the right equipment, it's not too difficult to do.

4.

In late 2004 a team of researchers from Johns Hopkins University and security vendor RSA Security attempted to replicate the encrypted challenge-and-response signal issued between a car and the average DST key. In practice, the auto manufacturers do not use all 1 trillion possible forty-bit codes; instead they use blocks of codes. That means that if you are part of an auto-theft ring and already have access to the manufacturer's code block, you're halfway there. But what if you don't have such access? What if you're Radko Soucek? Would you believe it is possible to "brute-force" (that is, try every possible combination within) a forty-bit ignition immobilizer DST from scratch, provided you have some additional computational resources?

With the cooperation of Texas Instruments (TI), one of a handful of vehicle transponder chip producers, the Johns Hopkins/RSA team set out to crack the encryption within five TI-produced DST tags whose corresponding code sequences were not provided.[51] In other words, the researchers had the locks but no keys to open them—a situation similar to that in which Tobias and Deviant Ollam find themselves.

Until their research, the TI algorithm for the forty-bit DST had been a secret. Yet, the Johns Hopkins/RSA research team was able to reverse engineer the algorithm within a few weeks. Right off they had a few advantages, among them time. The static part of the encryption process—the twenty-four-bit part of the forty-bit encryption that does not change—within the immobilizer chip had been developed in the early 1990s, back when computers were slower. By taking advantage of Moore's Law and using the faster computers available to them fourteen years later, the team found cracking that part of the code somewhat trivial.

Once the twenty-four-bit static part of the encryption was known, the researchers still needed to break the other sixteen bits. That requires more computing power, even today. As mentioned, forty-bit encryption can generate up to 1 trillion possible codes. A high-tech thief can't spend several days hacking your car's key fob sequence. The researchers had to design a system that was faster than today's commercial computers. The researchers custom-built a computer using field programmable gate arrays (FPGAs). An FPGA is the equivalent of several computers with linked motherboards whose different computer processors simultaneously handle one computation—like working through 1 trillion codes. (Note: The concept of FPGAs is important and will come up later in the book.)

The team used an FPGA board available for purchase on the Internet for about $200 and managed to crack thirty-two keys in parallel while running processors at 100 MHz. Had they used just one board, it would have taken them ten hours to crack one DST.[52] Instead the team bought sixteen FPGAs for around $3,500. Combining sixteen FPGAs into one unit allowed the researchers to generate all the possible random codes for each of the five DSTs in just under two hours. The researchers said later that they could distribute the computations differently and further reduce the processor hardware. "We think the entire attack could be done with a device the size of an iPod," said Adam Subblefield, one of the researchers.[53]

With the FPGAs the researchers had generated over 1 trillion possible DST codes, but which one would open the first of the five DSTs? The researchers had to reduce the time required to try all the codes they'd generated, so they used the brute-force method: They simply bombarded the five test DSTs with each of the possible codes until they found a match.

When they found a match, they did not stop; as mentioned previously, cars today use rolling or hopping sequences, so the researchers needed to find several matches in order to build up the rolling sequence. Once they had a few numbers that worked, the Johns Hopkins team wrote an algorithm that predicted the next code in the sequence for each of the five DSTs.

It took the Johns Hopkins/RSA researchers several weeks and thousands of dollars to design and build these systems, which critics say is a big obstacle for common street thieves.[54] However, once the initial work was done—say, by organized crime—word would spread on the streets to criminals like Soucek, who would download the algorithm from black market sites on the Internet with little effort. In other words, each criminal doesn't have go through this process; only one does.

The TI DST technology researched by Johns Hopkins/RSA is also used in the Exxon Mobil SpeedPass payment system. Instead of using a traditional credit card at an Exxon Mobil gas station, customers simply wave the contactless SpeedPass card in front of a gas pump. Like the vehicle immobilizer in a car, the pump queries the SpeedPass card; if the code is correct, the system allows the customer to pump gas, then bills the account for the amount. The Johns Hopkins/RSA team replicated theft at the gas pump; they managed to clone a SpeedPass card and pump gas using someone else's account (they even have video of it).[55] There is an important distinction however: The SpeedPass system has numerous layers of fraud protection on the back end, much like any credit card today. When suspicious usage occurs, the event is stored, and if it is deemed fraudulent, it is stopped.

That's not true with a car. Here the system is closed. There are no layers; there is no external monitoring of the key exchange between a legitimate and a fraudulent DST. A thief obtaining the proper sequence for your car renders it as vulnerable as any car without a vehicle immobilizer system. That said, a thief must still do other things to successfully steal the car, but he is most of the way there.

While the Johns Hopkins/RSA study only showed how to "hot-wire" a modern car and bypass the immobilizer security, another study, this one in 2007 from Ruhr University in Germany, showed how thieves could use a laptop to bypass the keyless entry system, eliminating the need to jimmy a door or break a car window to gain entry.[56]

The keyless entry system used by Honda, Toyota, Volvo, and Volkswagen, securing millions of cars worldwide and known as Keeloq, is manufactured by U.S.-based Microchip Technology. What's interesting is that both the vehicle immobilizer and the Keeloq keyless entry attacks are very similar. Because the algorithm behind the Keeloq keyless entry system was not published, flaws emerged only after the computing resources outside the company guessed them. "If [Keeloq] had made [their algorithm] public they would have found out 20 years ago that it's insecure," Timo Kasper, a PhD student who worked on the research, told Scotland's *Sunday Herald*.[57] "Now it's a little bit too late, because it's already built into all the garages and cars."

The Keeloq exploit, according to the researchers, uses a master key trick similar to that employed by Marc Weber Tobias. Because each car's individual keyless entry code is derived from its manufacturer's master encryption key (not a physical key), a criminal needs only to obtain the master key to derive the specific key for a given car. Someone with a little technical skill and the right equipment could sit in a crowded parking lot and wirelessly intercept countless keyless entry transmissions. With as few as two codes from the same auto manufacturer, a would-be attacker could use that pair to learn the master key. Once he'd obtained the master thirty-six-bit code, he could derive the unique code for a specific car.

The easiest attack scenario is the restaurant valet who copies both the keyless entry fob and the immobilizer chips while you eat, only to steal your car later that night. Keeloq has since changed the algorithm, but millions of cars with the older, more vulnerable code remain on the streets worldwide.

5.

So, should we throw out all our locks and, while we're at it, also throw our hands into the air? No. Even with their individual flaws, the technologies mentioned here do work—when used together. Layering slows a person down, said Deviant Ollam, long enough for a law enforcement response to arrive. "That's all security is, barriers to slow someone down. Even a safe, someone can make the best safe possible, but someone else will still get into it. I could thermite the safe. I might destroy what's in it, but I could get it with explosives let's say." The idea is to get the common criminal to realize it would take too much time or effort to break into this system, so they'd be better off moving on to a weaker target. Depending on what it is you are securing, the number of layers can be few to many.

"You'll hear people use terms like 'siloing,' 'onion layers,'" said Deviant Ollam, "but the idea is to have multiple phases to get people in, multiple authentication mechanisms. Not so many that it slows you down to nonproductivity, but if you have to punch a code and use a key, that's better than just a key, or just showing a badge."

Deviant divides all mechanical locks into four categories.[58] The weakest are common locks, which he simply calls basic. These are the file cabinet locks that can be opened with a paperclip and include combination locks like those used on gym lockers. Combination locks do not necessarily use pins, but the basic idea is the same. As the wheel is turned to a given number, a mechanism clicks into position inside the lock. Get the right combination, and the lock pops open. Some combination locks sold at the drugstore, said Deviant Ollam, can be opened using only a little metal shim cut from a beer can—with little art or science required.[59] Even without a shim, a combination lock can be opened by turning the wheel until you feel a slight tension on it. When you feel that, write down the number. There should be a dozen or so hits on any given lock. Now look at the numbers. Any numbers that fall between two numbers (i.e., 10.5, 31.5) you can discard. The remaining whole numbers will lead you to the final combination.[60] Basic locks, in Deviant's view, are those anyone could open within five minutes.

Next are what Deviant calls pick-resistant locks. These are sometimes branded as having an advanced keyway or labeled "contractor grade." These locks can be bought at hardware stores and sometimes include a mechanism that prevents very rudimentary attacks such as shimming. Sometimes they also have resistant pins to prevent manipulation or bumping. Bumping is on par with the simple opening of a combination lock described above; you insert a filed-down key and try to jiggle or bump the tumblers into their respective positions to open the lock. This involves a similar skill level as the jiggle key approach mentioned earlier and perhaps used by the car thieves who stole Brad Stone's Honda. Deviant said these pick-resistant locks are often sufficient for use in residential settings because common burglars do not pick locks on houses; such criminals break windows instead.

Following that are what Deviant calls high-security locks. Of course, many in the security industry bandy the phrase "high security" about, but Deviant has his own specific definition of what should constitute such a lock. Going beyond straightforward and typical lock designs, high-security locks (in his view) are augmented by additional mechanisms that, rather than merely providing greater manipulation resistance, require a skilled attacker, often operating with specialized tools and techniques, to open. Deviant estimates it would take an unskilled attacker more than a half hour to have any hope of compromising one. These locks can be purchased from any reputable locksmith.

Finally, there are what Deviant Ollam calls "unpickable locks," a rare breed in that they have no known attack vector or bypass weakness.[61] In this category Deviant Ollam currently places the ASSA Abloy Protec, which uses wheels; the ASSA Abloy Mul-T-Lock MT5, which uses specialized rotating disks; and the Evva MCS, which uses magnets and special notches in an innovative key design. These locks usually have to be purchased directly from the manufacturer.

Deviant suggests using high-security locks for external entry into a home or office and pick-resistant locks for internal rooms; inexpensive basic locks should never be used in commercial environments. If you have something sensitive to protect, then spend the cash and use an "unpickable lock" since part of what you are purchasing is peace of mind with the guarantee that no one has meddled in your affairs as long as the lock has not been physically destroyed.

Deviant echoed a refrain heard at many security conferences these days that the hardware world is still about ten to twenty years behind where the software world is today. The problems with vulnerability disclosure that we were dealing with in the software world are very much apparent in the hardware world. "You could say it is very in-house," said Deviant, "and it permeates this sort of theme of our knowledge is private and you don't belong here, you're not with us. That's all the way throughout the manufacturing industry."

Layering is also the basic idea behind recommendations for securing your car—high-tech antitheft system or not—according to *Edmunds*, the car magazine.[62] They recommend not leaving valuables out on the seat, having car alarms installed professionally, and never putting security alarm decals on car windows (it tells a thief exactly which system to defeat for entry). They say to consider using a hood lock to keep criminals from disabling a variety of security systems routed through the engine block. They go on to admonish you never to leave your car while it is idling, never to leave registration or other personal information in the glove compartment, and further to pay extra to park close to a stadium or other venue, arguing that the increased foot traffic, if not greater security, is worth it. The National Insurance Crime Bureau also has its own fact sheet full of similar tips.[63]

People who build cars, locks, and even cellular networks do a fine job (most cars are very reliable today, locks hold up well over the years, and calls don't get dropped as much as they once did), but most of these people lack security skills. Many of us are living with a false sense of security if we depend on one magic beep-beep to keep our homes, our cars, and our property secure. Sure, these are convenient gadgets, but the people who designed them don't often think outside the box.

They don't think like Radko Soucek.

CHAPTER TWO

The Dark Side of Convenience

The premium movie playing on Adam Laurie's hotel room TV screen may not necessarily be one he paid for, perhaps not one intended for his room at all. One night out of boredom, Laurie said, he became interested in his hotel room's TV remote handset and, in the process of exploring it, gained access to premium services, to other guests' accounts, and to the hotel's main billing server.

Unless they are accessing the Weather Channel or CNN, most people do not give the common hotel TV remote a second thought. Then again, most people are not Adam Laurie. He is the chief security officer and director of a London-based networking company called Bunker Secure Hosting, housed inside a decommissioned missile silo outside of the town of Kent.[1] His frequent travels and speaking engagements are the result of Laurie's world-renowned expertise in wireless vulnerabilities found in many gadgets today, including hotel TV remote systems.

Laurie, who still uses the nickname "Major Malfunction," discovered the possibilities after idly tinkering with infrared codes via his laptop one night in a Holiday Inn hotel room.[2] Setting down his laptop, Laurie said he wanted to retrieve a cold beer from inside his previously unlocked minibar. Somehow he'd managed to change one critical value via the TV and locked the minirefrigerator. If only to rescue his beer, Laurie said he was compelled to rediscover the exact numeric value that would unlock it. And, of course, one thing led to another.

Infrared signals on consumer gadgets are easily overlooked ("security by obscurity"). By comparison, there's the very basic radio frequency controls used in garage door openers. Garage door openers can be manually configured via a dipswitch circuit with eight possible on-off positions. That leaves 256 possible code combinations. Laurie has demonstrated at various security conferences a script he created that can run through all 256 combinations in a matter of minutes. With the script on his Linux laptop and a radio antenna, he can open just about any garage door. (He has also used a variation on the keyless entry attack we discussed in the previous chapter to lock an employee's car in the parking lot after the owner attempted to unlock it. In Laurie's telling, the employee couldn't figure out why his key fob wouldn't open the door, much to the amusement of the rest of the staff watching from a nearby window.)

With TV remotes very few industry standards exist for infrared television remote signals. Those that do are proprietary. For example, a Sony TV remote won't work on a Samsung TV but might work with another Sony product, such as a Sony DVD player. No encryption or authentication is required to use a remote. No authentication handshake says that only a Sony remote with gadget number x can connect to a TV with gadget number y. This gives us the convenience of universal remotes, even though they require some initial programming by the end user if only to tell the universal remote what proprietary code to use.

Unlike the home version, hotel TV remotes include additional groups of code. The home edition includes volume, channel select, and text mode. The hotel version includes codes for "alarm clock," "pay TV," "checkout," and "administration" (such as housekeeping). Hotels, however, use an inverted security model in which the end gadget, in this case the TV, filters the content.[3] In other words, premium movies are broadcast all the time; you just need a way to access them.[4] Instead of residing in a central server, access control is literally in the hands of paying hotel room occupants—whether they realize it or not.

Laurie found he only needed a computer running the Linux operating system, an infrared transmitter, and a USB TV tuner to access these extra groups of codes.[5] While staying at a Hilton Hotel in Paris, he automated his attack, which enabled him to snap photographs of the various channels he could see and manipulate.[6]

If he'd had malicious intent, Laurie could have zeroed his minibar balance, watched free premium movies, or surfed other people's e-mail.[7] Instead Laurie decided to deface the hotel welcome screen, take a photo, then restore the screen to its previous condition, later using the photo to show the hotel staff what he'd been able to do. "If the system was designed properly," Laurie said, "I shouldn't be able to do what I can do."[8]

Yet, the ability to access the minibar records through the hotel television shouldn't be too surprising. Hotel TVs are connected by coaxial cables to a little metal box. So are the room's premium TV channels, VoIP, minibar, and Game Boy or Wii entertainment system. This bundling of premium services is convenient from a hotel's point of view; management doesn't have to rewire each room whenever it adds a new service. And it's convenient from guests' perspective; they can check out anytime and bypass the front desk. But there's a flaw in all this convenience.

Using a computer TV tuner and his laptop keyboard as a remote control, Laurie said he is able to access information intended for other rooms within the hotel.[9] Thus, Laurie can change the code and see the billing information for another room, any Web mail that person might be reading, or whatever premium porn channels that guest might be watching at the moment.

The hotel assumes that only you can see your account information. It further assumes that most people aren't connecting their laptop computers to their room TVs and accessing the hotel's private configuration codes. For the most part, that's true. But Laurie isn't the only security researcher to publicize this particular design flaw. Paul "Pablo" Holman of the Shmoo Group has also gone public with his own findings on hacking hotel room TV remotes.[10] Another security researcher used only a basic cable converter box purchased on eBay to intercept the hotel codes in his room.[11] Others have also found additional ways to defeat hotel TVs, such as plugging the coaxial cable into their laptops.[12]

Realizing that his otherwise trivial hotel TV remote more or less holds the keys to the entire kingdom, Laurie has experimented at other hotels. He's seen only three or four different back-end systems used. By one estimate, one of up to 16,000 possible code combinations is required to unlock the services on any given hotel system; each new location could take Laurie hours to decipher.[13] To speed that process, he created an automated script like the one he used to crack garage door openers to divine a particular hotel's relevant codes in about a half an hour. Laurie has no plans to release that script to the public. It exists only to further satisfy his curiosity.

In one hotel Laurie inserted an unblocked porn channel image onto the background of the welcome page—temporarily and only to show the executive staff. Similarly, he once accessed the hotel's main server and had the option of crashing the entire system—if he wanted.[14]

That's where Laurie as a researcher differs from the criminals sometimes referred to in the media as "hackers." Laurie uses his experiences to educate people about the dark side of common gadgets. But what if someone really wanted to be malicious? Could that person use a common gadget to get sensitive information about us?

Lack of authentication allowed Laurie to gain access where he should not. In many systems we take for granted, this lack of authentication is all too common because the designers have not thought through the various ways in which someone could attack. As we will see, our growing need for convenience makes us accept clever shortcuts in exchange for security, shortcuts that in some cases may cost us money or, in extreme cases, our lives.

1.

Depending on whom you ask, the U.S. electrical grid is either totally secure against, or completely vulnerable to, outside attack. The grid comprises individual gadgets (pumps, sensors, switches) that are linked via

Supervisory Control and Data Acquisition (SCADA) systems. SCADA systems control the distribution of electrical power and are both wired and wireless. They use their own network, not the Internet.[15]

In the past, SCADA networks were mostly unprotected, consisting of individual gadgets strung together with little care for authentication of legitimate users or encryption. During the early 2000s, however, shortly after 9/11, the North American Electric Reliability Corporation (NERC) made a series of changes, including requiring energy providers to use virtual private networks, which allow for user authentication, and Advanced Encryption Standard encryption. In addition, the utilities' SCADA systems needed to use firewalls. The 2007 Energy Independence and Security Act gave utilities until 2009 to comply, or NERC would begin to assess fees.[16] By April 2009, however, "most of the electric industry had not completed the recommended mitigations," said Rep. Bennie G. Thompson, chairman of the U.S. House Committee on Homeland Security, "despite being advised to do so by the Federal Energy Regulatory Commission and the North American Electric Reliability Corporation."[17]

Yet, there still seemed to be a national urgency. In late 2009 nightmare scenarios in which hostile foreign powers access these SCADA gadgets and disable them, causing widespread and prolonged blackouts (perhaps as part of a larger attack), were presented as accomplished fact. The segment "Sabotaging the System" appeared on *60 Minutes* during the hot TV ratings period in November 2009.[18] It quoted six U.S. government officials attributing a two-day blackout that affected 3 million people in Espirito, Brazil, to the work of cyber-criminals, with an implicit warning that such an event could be repeated elsewhere. Additionally, the U.S. government officials linked another blackout in January 2005 in Rio de Janeiro to the handiwork of cybercriminals.

In reality, the Espirito, Brazil, blackout was caused by poor maintenance—specifically, sooty insulators on the high-transmission power lines.[19] Brazilian government officials noted that a drought had created deposits of dust and soot, most likely from a brush fire in the Campos region of Espirito Santo.[20] This, they said, was the likely cause of the blackout. Brazilian officials have further denied U.S. reports that their SCADA power system is connected to the Internet.[21]

This hasn't stopped U.S. officials from alleging that cybercriminals have brought down SCADA systems. Richard Clarke, former special adviser on cybersecurity to President George W. Bush, also named Brazil as a cyberblackout victim, and in a 2009 interview with Wired.com, Clarke echoed a charge by Tom Donahue, the CIA's chief cybersecurity officer, that parts of the U.S. system had also been taken down by cybercriminals.[22] Despite these vague comments from government officials, there has been no specific published example of someone compromising a U.S. power grid SCADA system.

Complicating the discussion about the relative security of SCADA systems is an international movement to monitor individual energy use via electronic smart meters in the home. The United States has fallen behind Asia and Europe in creating these so-called smart grids.

Unlike SCADA systems, which are primarily designed to distribute power in bulk, smart grids are designed to fine-tune energy usage at the individual level. A critical component of the smart grid is the gadget known as a smart meter, which allows for automatic meter readings (AMRs) within the home. These gadgets can also connect to individual appliances, such as a dishwasher, relaying information back to the utility and to the consumer about the energy costs associated with its use. In Japan, where AMRs have been in use for a few years, in-house batteries are also used so that customers can buy electricity at night when demand, thus rates, are lower, then use the battery during the day when rates are higher. Solar homes in the United States currently use batteries to store electricity and sell it back to the grid, and savvy customers can do this during peak day usage when the prices are high. So, home owners with smart meters could audit their appliances and perhaps save money by time shifting use of these appliances to periods when local energy was least expensive (overnight), or they could simply replace inefficient dishwasher models.

To catch up with the rest of the world, President Barack Obama's 2009 stimulus package, the American Recovery and Reinvestment Act, included financial incentives to U.S. utilities to install smart meters in every home by 2012. In response, in late 2009 and 2010, utilities around the United States began installing these gadgets, with one estimate holding that more than half of U.S. households would be wired with them by the end of 2010.[23]

But there are problems. "With the advancement of the 'Smart Grid' and AMR systems, without the proper security precautions, the electric grid is now more vulnerable than ever," said Jonathan Pollet, founder of Red Tiger Security.[24]

Speaking at Black Hat USA 2010, Pollet said that his company conducts penetration tests for utilities in which he and his team, with the permission of the utility, attempt to hack their way inside. After one hundred such assignments, his company logged over 38,000 software vulnerabilities in the operating systems used on the SCADA networks. He also found a number of personal, commercially available software applications such as BitTorrent on these systems; the personal software may have vulnerabilities as well. While not serious enough to cripple the SCADA systems, a flaw within any piece of the commercial software, such as a buffer overflow, could provide someone outside the network with access to the larger power grid. The smart meters in the home, according to Pollet, provide yet another vector for a possible attack on the U.S. electrical system.[25]

In another session at the same Black Hat conference, Shawn Moyer and Nathan Keltner, researchers at FishNet Security, showed the audience a proof-of-concept smart grid attack scenario.[26] Using radio equipment and open-source software, they located the signatures of smart meters on an open network and circumvented the encryption used, allowing them direct access to the meter. An attacker could do this and issue commands that would either jack up some people's rates or shut off power to others.[27]

"Although the security initiatives and elected officials have good intentions, they have missed the window of opportunity to truly integrate security from the beginning by several years," wrote Tony Flick of FYRM Associates in notes preceding his presentation titled "Hacking the Smart Grid" at Black Hat USA 2009. "Similar to the credit card industry, banking industry, health care industry, and most industries that conduct business online, the next electrical infrastructure will feature security as an add-on that is applied after the smart grid is implemented."[28]

Smart meters, unlike desktop systems, embed software programming within a chip; in other words, the software code doesn't sit in volatile random-access memory but on a nonvolatile read-only memory (ROM) chip. This "tamperproof technology" is used for smart cards, such as those used for boarding a bus or train, and for medical gadgets, such as personal insulin pumps. Embedded systems are efficient because a coded chip requires less maintenance for something "out in the field" than a full-blown operating system. But "nonvolatile" doesn't mean "invincible": The ROM can still be "flashed"—that is, its code can be rewritten remotely. And this, researchers have found, is true with most smart meters on the market today.

With smart meters, attacks such as unauthorized software updates can affect millions of users at once. Instead of attacking the larger, older SCADA networks, an attacker might only focus on the newer smart meters hastily installed. Flick points out that Austin Electricity in Texas began installing smart meters in 2002, and another project commenced at Salt River in Arizona in 2006. Both installations were completed well before the initial studies by NERC and the National Institute of Standards and Technology and might be more vulnerable than units installed in 2010.[29]

The principal flaw, say researchers, is that the smart meter units are designed to communicate with the utilities and with each other. A computer worm in one could propagate exploitation of a flaw to millions of interconnected units. And that's exactly what another researcher showed at Black Hat in the summer of 2009. Mike Davis of IOActive found that he could rewrite the firmware of one smart meter and propagate a worm across millions of similar units.[30] Davis told *Internet News* that he could also rewrite the code for more than one smart meter. "Due to the peer-to-peer nature of the network, we could hop from one meter to the next updating the firmware," he said, "so that essentially they could all be running a custom firmware patch that any attacker could use to insert into the network."[31]

"We can switch off hundreds of thousands of homes potentially at the same time," Davis told *The Register*. "That starts providing problems that the power company may not be able to gracefully deal with."[32]

Fortunately, researchers like Pollet, Flick, and Davis have all shared their information with the smart meter manufacturers, some of which are making changes to their security designs. Other manufacturers insist it is up to the utilities to configure security on their own.

This is neither the first nor the last time that gadgets designed for our convenience have been hastily implemented in the field without enough security. In this case, the costs of security mistakes will be huge. A critical feature of the AMR is the ability for a utility to shut off a delinquent account holder. At present, a utility worker must visit the residence or business; in the future, with smart meters, this will be done remotely. Now, imagine what might happen if Davis's worm triggered the off switch on millions of smart meters. It would make the blackout in Espirito, Brazil, look small by comparison.

2.

When we empower gadgets to talk to each other and to the Internet, yet haven't take the first step to secure them, we're inviting trouble. This concerns Howard Schmidt.

One sunny spring afternoon a few months before his appointment as President Obama's White House cybersecurity coordinator in 2009, Schmidt told me about a new Blu-ray DVD player he and his wife had just purchased at Fry's Electronics. In addition to having several USB ports, Schmidt's new DVD player, like many TVs today, also had its own IP address. "They're great gadgets," Schmidt said of the Internet-enabled DVD player and TVs. "But what are [the vendors] doing about security?"[33]

The convenience is obvious. The particular player Schmidt bought comes enabled with Netflix and Pandora, two popular Internet services. Without using a computer, Schmidt and his wife can stream the latest movies and music on their TV. Schmidt, who also served as vice president, chief information security officer, and chief security strategist for eBay, also understands the potential of connecting a DVD player to the Internet to buy something online while watching QVC. Rather than waiting for a telephone operator, you can punch an order code with your TV remote control and make an instant purchase.

Admittedly, being able to pause a DVD and then go onto the Internet to look up an actress's name or prior roles is useful. Schmidt suggested entertainment companies might also make additional content available on the Internet, such as scenes from a director's cut, an alternative ending, or updates to the material in a documentary.

A better use for Internet-connected TVs might be the Emergency Broadcast System (EBS), said Schmidt. Introduced on August 5, 1963, EBS was designed to give the president immediate access to the nation during the Cold War. It included harsh tones to attract attention to the television set or radio. After a few seconds of beeps, the station would then broadcast an emergency message direct from Washington, DC. The system has never been used on a national level, so most of us have only heard the test message required by the Federal Communications Commission (FCC), but the system has been used several thousand times to broadcast local severe-weather announcements in the midwestern and southeastern United States.

Schmidt outlined to me a fictional scenario in which Seattle's Mount Rainier, a semiactive volcano located in Washington State's Cascade Mountains, blew its top. Whereas the old EBS might be able to alert perhaps a few million people in the surrounding area via TV or radio, an Internet-aware television entertainment system might reach several million more and could cut to a live feed from the local emergency command center. Such an enhanced system could provide actionable information, perhaps displaying updated evacuation routes and live feeds from the ground. This could also work for hurricane or tsunami evacuations, terrorist attacks, or even AMBER Alerts.

But beyond the positives, a threat that isn't quickly realized also exists. Adding functionality to existing technology makes for some unintended consequences.

It is important to note, first, however that gadgets not directly connected to the Internet might also pose a risk. Slingbox is a device that can broadcast signals from your TV onto any computer or mobile phone. In 2007 a group of researchers from the University of Washington found they could capture enough broadcast information from early versions of Slingbox to determine what show or movie a person was watching in the next room.[34] They could predict the title of a video being watched 85 percent of the time. While this could be effective in blocking access to certain content by minors ("Johnny, turn off that video"), the ability to decipher the video stream without the viewer's knowledge could also be embarrassing for some adults. Slingbox has since added Secure Socket Layer encryption to its video streaming sessions, and we'll discuss the weaknesses in various other forms of wireless transmissions in the next chapter.

Ultimately, anytime you connect a gadget to the Internet or otherwise broadcast a signal, you increase the risk of data leakage or failure (a denial-of-service attack). It's one thing when you're talking about not being able to watch a movie at home or make toast. It is quite another when you're talking about any number of gadgets related to health care.

Along with the smart grid incentives in President Obama's 2009 stimulus bill was the HITECH Act, which offers financial incentives to health-care companies that create electronic health records and the gadgets to support them. The bill also gives financial incentives to doctors, hospitals, and health-care providers to convert their paper medical records to this new digital format. Except there is no agreed-upon digital format and no existing digital medical standard, let alone any established set of accompanying security safeguards. So, there's a very high potential for early adopters to develop leaky data systems. Thus, the HITECH Act also contains high financial penalties for data breaches affecting health care.

The fundamental concept behind the HITECH Act is good: You could check into any hospital anywhere, and doctors there would have immediate access to your complete medical history. In cases where you lost consciousness, electronic health records would be ideal, providing the medical team with much needed medical history, leading to an accurate diagnosis. On the other hand, as we have seen, in the rush to "go online" or generally provide convenience, matters such as authentication (the ability to positively identify the party) and encryption (the ability to obfuscate the data) are omitted. On the continuum of personal data that can be leaked, health-care information is perhaps the worse. Unlike with financial fraud, if someone uses your medical insurance to pay for his own operation, that operation remains a part of your medical history. It could influence future health-care decisions, treatments, and insurance rates.[35]

Congress has left out health-care security details before. In 1996, the Health Insurance Portability and Accountability Act (HIPAA) stated that health-care companies could not track patients using their Social Security numbers. This was a good idea, but there was little enforcement. Thus, more than a dozen years later, my own doctor's office had a space on its patient information form asking for individuals' Social Security numbers. When I asked why, in the age of HIPAA, that was still there, I was told, "It's always been on the form." I left the space blank, but how many others have not?

HIPAA anticipated the type of e-health innovation called for in HITECH but failed to address it. After more than a decade, HIPAA technology subgroups have not produced a substantial body of standards around common medical needs such as image file formats for X-rays, CAT scans, and MRIs. With HITECH, Schmidt sees the opposite problem. This time, there's too much of a rush, hastened by the financial incentives to go online early rather than when the system is ready. "I look forward to going electronic," said Schmidt, "but I have a tremendous amount of concern about building a really, really good healthcare infrastructure and then securing it later."[36]

In this rush to adopt, a lot of medical facilities will use existing off-the-shelf technologies—for example, doctors will use tablet PCs to connect remotely to hospital databases. Here, the serious security flaw might reside not within the software they use for an image file but in secondary Wi-Fi protocols and how they are implemented at the hospital.

Schmidt has advised Congress that a digitized health-care system should include strong authentication and encryption (the ability to obscure the data so that others cannot eavesdrop) and "not just build a really cool health-care system so a doctor can pick up his BlackBerry and say, 'Yes, Howard Schmidt, here's his patient data.'"

Since such a system will be wireless, Schmidt said that something else must be sitting out there doing all that security. "It's got to be part of the requirements," he said, "and not just an optional thing."

In one of his past jobs, Schmidt was the head of software security at Microsoft and involved in the infamous Trustworthy Computing initiative proposed by Bill Gates in 2002. Trustworthy Computing established a process known as Secure Software Development Life Cycle with various gating factors that allowed concerns over security to put an entire project on hold.

"Remember how Microsoft got criticized because Vista slipped [its ship date]?" Schmidt reminded me. "Well, in the background they were trying to fix security vulnerabilities which we would criticize if they had let it ship." In other words, having a timetable is nice, but it must contain some stipulation that the timetable will slip if a particular privacy or security requirement is not met. HITECH contains no such provisions.

The emphasis on convenience is creating a similar lack of concern for privacy and security in the home. In the past, DVD players were not networked in the home environment beyond the TV and maybe a stereo receiver for home theater speakers. Software vulnerabilities didn't really matter within any of these individual gadgets, as all of them were connected to the same local network: your home. Connecting one or more of these to the Internet gives some individual on the other side of the world the ability to reach inside your house and exploit that previously harmless software vulnerability and access the controls of a DVD player sitting in your living room. More likely, the person could create a denial-of-service attack by changing the internal programming, preventing you from seeing a show you want to watch.

"I don't want someone sitting there burning a new version of firmware onto my DVD player that's going to render it unusable." Schmidt said. "Think of the expense you'd have to go through trying to convince the manufacturer that, hey, there's a problem that's not really their problem."

Maybe your Internet-enabled DVD player isn't connected to anything else in the home. But maybe the TV is connected to your computer and, from there, to your home computer network. Now someone from some other country could seriously muck around with your DVD player and also access personal data from your computer or home network.

3.

How serious is the risk when health-care systems are connected to the Internet? "Very serious," said Benjamin Jun, chief technical officer of Research Cryptography in San Francisco. If cybercriminals can attack embedded systems in DVD players, they can also attack more critical systems, such as medical gadgets, where changes made by outsiders can have life-or-death consequences.

Implantable medical devices (IMDs), such as pacemakers or implantable cardiac defibrillators, are lifesavers, allowing recipients to lead normal lives. However, these internal gadgets need to be monitored and occasionally recalibrated. Presently this is done at a doctor's office. For some, getting to the doctor is a challenge, so a few health-care gadgets include IP addresses that allow medical personnel in another city to monitor them over the Internet and make necessary adjustments remotely. For patients in rural North Dakota, that's convenient; they don't need to make the hour-long trip into town for ongoing medical care. But, as Schmidt has pointed out, having another IP address to contend with creates problems; most people won't know how to configure firewalls specifically for their medical gadgets.

Like Schmidt, who talks about the potential for someone to reconfigure the firmware on his DVD player, Jun looks at ways in which someone might reprogram embedded systems—systems that are too small to run traditional software, so the software is baked into the chip.

Jun said that hardware today is basically free; what you are paying for, what differentiates most gadgets, is the software. Without the software, the early Apple iPod was basically a large flash drive; software for iTunes gives that gadget the most value. So, if you can alter the embedded software, get your gadget to do things you haven't really paid for, then that's interesting. It's also a huge problem for manufacturers who license and charge fees for product features separately.

Jun cited one example: medical glucose tests. Go to any drugstore, pay US$20, and you can get an electronic gadget that works with a particular brand of test strips. The electronics inside the gadget cost much more than $20, of course, but the manufacturer is willing to take that loss up front. The company makes its profit through repeat sales of its branded test strips, which it relies on selling at a slightly higher price. "In the case of the glucose test meter, it's how the strips are actually calibrated, how they interact with the system, that gives the system value," said Jun.[37]

But the glucose test strips are being counterfeited. Inexpensive test strips that have not been calibrated or authorized are being engineered and sold at lower cost to operate with these proprietary systems. They work because neither the legitimate nor the counterfeit test strips are authenticated. The danger here is that counterfeit test strips could give false results.

Obviously the manufacturer's challenge is to make its system reject these counterfeit strips. Jun suggests adding a security chip to the test strip, which would issue a challenge and response, use a digital certificate, or employ some other authentication scheme. If the insulin pump doesn't recognize it, the strip will be rejected.[38]

Many different industries are facing essentially the same problem. For digital content, it's the authorization to play only a game or movie that has been paid for. For a mobile phone, it's the authorization to use a particular cellular carrier's network. And for implantable medical gadgets like pacemakers, it's authorization to access the device. But with IMDs there is an additional problem when introducing security policy controls that are too strict.

Imagine a patient entering a developing country's emergency room alone and unconscious. The staff determines that he has a number of IMDs. Perhaps the staff also makes lifesaving adjustments in order to treat him. They learn about the patient's history from the IMD and perhaps obtain his full name and address. This is the success story.

In a second scenario, the patient controls all the information on his IMD. Use of strong cryptography limits unauthorized exposure of data, which is good in most cases. But remember that in this case the patient is alone and unconscious. In the second scenario, the patient might die a "John Doe."

This scenario might be avoided through the use of medical smart cards uniquely tied to a doctor's identity. This third scenario would require some kind of worldwide electronic health standard. Once authenticated, that doctor could then control the IMDs, perhaps read the patient's health history, and maybe save the patient's life.

Another possibility might be to encrypt a key onto a card or medical ID bracelet worn by the patient, allowing medical personnel to use the key to unlock the IMDs when necessary. However, the person might leave the card at home or lose the bracelet. Additionally, the encryption key could be broken and used by nonmedical personnel.

To highlight security problems with IMDs, researchers from Beth Israel Deaconess Medical Center, Harvard Medical School, the University of Massachusetts–Amherst, and the University of Washington broadcast a remote malicious code directly into an Internet-enabled pacemaker.[39] In this case, because the gadget regulates the heart, the researchers, whose work can be found on the Secure Medicine website,[40] didn't need to reprogram the pacemaker or install special executable instructions to create harm. They only needed to corrupt or interrupt the existing code to produce a catastrophic result: death.[41] This would be a denial-of-service attack on a human.

The research is quite credible. One of the researchers, Tadayoshi Kohno, was previously involved in the project that showed how Slingbox signals can reveal what a person is watching on his computer. And other members of the research team, Thomas S. Heydt-Benjamin, Hee-Jin Chae, Benessa Defend, and Kevin Fu, had previously studied the security of electronic smart card–based public transit systems, which we talk more about in Chapter 5.

These researchers sent enough gibberish code for a Medtronic-brand pacemaker to become erratic. This, in turn, would lead to an irregular heartbeat, which would cause arrhythmia. In some cases, if emergency care is not obtained soon enough, arrhythmia alone would be enough to kill someone.

Prior to publishing that study, the researchers published a list of suggested mitigations for such an attack.[42] These mitigations include

- limiting the use of "test mode" in a pacemaker
- limiting broadcasts from these gadgets
- limiting access to the gadget ID itself

The pacemaker manufacturer, Medtronic, wasn't convinced. In a published response to the Secure Medicine researchers' work, Medtronic cited a number of reasonable counterarguments.[43] For example, someone would have to know the IP address of the gadget. This could be learned, however, through a data breach of medical data from a hospital or doctor's office. Even if there wasn't a targeted data breach at a health-care facility, security by obscurity is not enough to keep you safe from a random pacemaker attack. The pacemaker companies bank on the fact that no one will have the Internet address for these gadgets. But that simply isn't the case.

In 2009 researchers at Columbia University began a survey of all the unprotected systems connected to the Internet, which would include remotely monitored medical gadgets.[44] The team, sponsored by the Defense Advanced Research Projects Agency (DARPA), the Department of Homeland Security, and other government agencies, used a scanner script to send a password for common gadgets, like home network routers or Schmidt's DVD player. The researchers were concerned that gadgets using default passwords are vulnerable to remote operation.

"People tend to buy stuff and bring them to work and just plug them in," said Salvadore Stolfo, the Columbia University professor overseeing the research. "So we think we'll be able to find vulnerable devices in highly sensitive places."[45]

For example, the team might send "admin," the default administration password used by the popular Cisco Linksys router, and connect to someone's home router. If the gadget responds with a command prompt, the research team then ends the connection session, records the event, and moves on. The event is also recorded by the gadget's Internet service provider, so the team leaves behind a link to a page explaining the project and suggesting ways that the gadget owner can—assuming he reads his log files—"opt out" of the study in the future and mitigate the chances of another party's also gaining access to the gadget. Extrapolating from the early numbers, the research team speculates that by the end of 2010, they may find as many as six million "stealth" gadgets that are otherwise accessible from the Internet, some of which could be implantable medical devices.[46]

Medtronic points out that interrupting a medical gadget would require specialized equipment and considerable resources. The researcher's experiment required more than $30,000 worth of lab equipment and a continued effort by a team of specialists from the Universities of Washington and Massachusetts.[47] However, foreign governments or organized nation-states might fund such an endeavor if the heart-patient target were, say, a political figure. What if the target hospital were Bethesda Navy Hospital and the attacker had access to the records of a president, vice president, or member of Congress? If the target were the chief executive officer of a major corporation, a ransom could pay for the initial resources and more.

Finally, Medtronic points out that the programming for its gadgets, like that for keyless entry systems and smart meters, is proprietary, usually a secret. But Jun, Schmidt, and other security researchers point out that we needn't wait until the bad guys attack these systems. Within the tight confines of embedded programming, potentially significant layers of security protection are already available. Jun said it makes no sense for the gadgets not to encrypt data, yet link to the Internet—they lack commonsense protections already found on your desktop computer at home. (Chapter 7 discusses some associations that are creating new security standards specifically for medical gadgets.)

Again, if there's a financial incentive, attacking a pacemaker might happen sooner rather than later. Here, the goal isn't to inject or run code. In this regard, Medtronic argues that the amount of code within the pacemaker is too limited. Again, that's true, but the attacker's goal is only to destabilize the pacemaker, not to reprogram it. The attacker only needs to introduce enough garbage code that the gadget will fail to work properly. And that is possible to do.

4.

Consider the tale of one ordinary shipping container at a major port. Like the other containers in the shipyard, this one is marked with an active RFID transponder similar to that in auto antitheft systems described in Chapter 1. This one contains a wire antenna, a tiny embedded radio transmitter, and a battery that broadcasts a short string of data—in this case only 256 bytes. Within this particular 256 bytes, however, is a short string of code with the potential to disrupt the entire port.

A Structured Query Language (SQL) injection is an attack in which a piece of special code disrupts an ordinary SQL database (a common type of database used for a wide variety of data) by making use of special characters. Websites are vulnerable to this attack if they don't filter out special characters in user-input fields. RFID hosts do not filter input or authenticate the RFID tags, and they are therefore vulnerable to these same attacks, according to researchers in Amsterdam.[48] As with the string of code we just said could disrupt a pacemaker remotely, here the 256 bytes in the RFID tag need only to disrupt the flow of information within a large and very critical database like that at a major shipping port.

SQL injection appends a string of code where code isn't necessarily wanted. In this case the SQL injection will try to pass a confusing request to the shipyard's SQL database, which is presumably connected to other databases or even the Internet itself. The database will try to execute the string, which will cause it to crash. Perhaps the entire port system will be compromised. By then the container will be long gone, anonymous among the thousands of other containers in the yard, waiting to be washed, refilled, and sent somewhere else. The malicious code (now a part of the RFID ecosystem) will be imprinted on other RFID tags on other containers, spreading the system-corrupting infection throughout the seaport and perhaps the world.[49]

While this may sound impossible outside the plot of a techno-thriller, unfortunately it is not. When I first wrote about this in 2007, I heard from the Smart Card Alliance, an industry lobbying organization, which argued the above scenario hasn't yet happened in the real world. That's technically true. In a paper titled "Is Your Cat Infected with a Computer Virus?" researchers from Amsterdam only looked at the practice of injecting domestic animals with RFID tags for easy identification and access to veterinary records.[50] They wondered whether a maliciously coded animal RFID tag could be used to corrupt a veterinary database. The same logic, however, could be applied to tags currently used on shipping containers or on products on the shelves of any big-box store.

Specifically, the team of researchers at the Vrije Universiteit in Amsterdam asked why host systems should implicitly trust input from any RFID tag. In other words, they challenged the underlying argument that these large database systems do not need to defend against malicious code attacks from RFID tags. They found most systems did not authenticate the RFID transponder. (Chapter 5 discusses RFID systems in greater detail.) So what happens when that data is garbage?

Indeed, prior to the paper's release, most people never expected to find SQL injection and buffer overflows—two traditional methods of attack on the Internet—within the mere 256 to 1,024 bytes found in most RFID tags. Yet, the researchers, thinking outside the box (which is how cybercriminals typically think), created just that. Technically what they did is called fuzzing, where you simulate code to trigger a buffer overflow or a cross-site scripting attack. The researchers realized that if just one infected RFID tag could be put into circulation, that infection could spread far and wide. Thus, something very simple, an RFID tag, has become very complex and dangerous.

5.

Perhaps the most technologically complex gadget we own is parked inside our garages. By one estimate, the typical car has 100 MB distributed randomly among fifty to seventy little computers, or electronic control units (ECUs).[51] The first ECUs resulted from actions taken in the 1960s and 1970s by California's Air Resources Board. Today's ECUs are given specific tasks, such as controlling the antilock brakes, the lights, the volume of the multimedia player, warning signals, and the flow of fuel to the engine.[52] Features that we consider convenient require the ECUs to communicate with more than one system, and a controller-area network gives the car's ECUs the ability to communicate with each other. For example, electronic stability control requires communication between the accelerator, the brakes, and the individual wheels.[53]

A group of researchers who decided to see whether an adversary could take over a car's central computer system was shocked to discover the ease with which it could disrupt the whole system. (That said, I should stress that the risk to your car and my car remains quite low. For example, an adversary would have to gain physical control over the car's electronics through a tiny port.) Since 1996, all U.S. cars have been federally mandated to have an onboard diagnostics port designed to give auto mechanics, not just dealerships, access to the data systems within the cars. Unfortunately, that same port might also allow a cybercriminal access as well. However, the attack scenario requires that an attacker spend a moment reaching under the steering column where the port is located and patch in code that could later have serious consequences for the driver. Again, the parking lot valet might be a good suspect.

Projecting forward, however, researcher Stefan Savage, from the University of California, San Diego, said the real concern is with wireless sensors on newer cars. Indeed, a separate research project used the tire-pressure monitoring system (TPMS) to gain access to a car's ECU.[54] In 2000, when low pressure in Firestone tires began causing accidents in Fords, Congress enacted the Transportation Recall Enhancement, Accountability and Documentation Act, which led to the creation of TPMS wireless sensors to warn of low tire pressure. The researchers at Rutgers University and the University of South Carolina used the wireless sensors to control the low-pressure warning lights on the dashboard of a car traveling at highway speeds more than one hundred feet away. These wireless tire-pressure monitors don't use authentication or validate input of new data, so rewriting is possible.

What these researchers did—trigger the tire-pressure warning system—may not sound like much, but car manufacturers are building more and more wireless systems into cars. Increasingly researchers are sounding a warning about the ultimate consequences. And the auto industry appears to be listening.

One manufacturer, Ford, is thinking ahead. Ford has partnered with Microsoft to create the SYNC system, allowing drivers to wirelessly control their music selections and use voice commands. In 2011 models of the Ford Edge and MKX Lincoln and the 2012 Ford Focus, the second-generation SYNC technology will update onboard firewalls to protect the wireless network as well as the ECUs. Additionally, second-generation SYNC will include antimalware technology and use Wi-Fi Protected Access 2 for encrypted wireless access.[55]

Savage cited an analogy, saying the desktop PC did not have so many problems until it was hooked up to the Internet. The same could be said for the modern car. For convenience, auto manufacturers have added hundreds of little features that range from the useful (governing the flow of fuel into the engine) to the luxurious (remembering preferred seat positions for different drivers). We've added considerable complexity to cars we once could service in our own garages.

In 2004, after Mercedes placed fifteenth in J. D. Power's annual Initial Quality Study with 132 defects found per one hundred cars, the company admitted it had gone too far in adding gadgets and removed several hundred electronic features from its models the following year. While introducing the new 2005 line of cars, Stephan Wolfsried, Mercedes-Benz's vice president for electrical and electronics and chassis development, said, "Last year we removed over 600 functions from our cars—functions that no one really needed and no one knew how to use. It is our aim to focus on things that make sense. The driver will not have the option to adjust everything that is adjustable."[56]

With all those features, only one needs to be vulnerable. That's because the automobile industry never put much thought into designing a unified operating system or platform to tie all the electronic subsystems together; manufacturers just strung together disparate features with very rudimentary security. Worse, one can compromise the car and completely own it; there's no back-end system to detect the attack and stop it.

Savage's research team from the University of California, San Diego, and the University of Washington created a program known as Car-Shark that allowed them to throw random code at their test car and probe its weakest links.[57] As in the case of bringing a shipping port to a dead halt or causing a pacemaker to beat erratically, the typical outside attack on a car would use gibberish.

Using CarShark, the researchers discovered codes that would increase the volume on the radio and prevent the driver from resetting it. They could broadcast various messages on the radio display, such as "I PWN YOU."[58] They could also produce various warning chimes and signal clicks randomly over the radio speakers. Those are just pranks, however.[59]

More ominously, they could falsify readings from the fuel gauge and speedometer, disable the antilock brakes, selectively brake individual wheels on demand, and even stop the engine. The researchers found that they could do this even while the car was speeding down a highway.

The cars in the research project—indeed, like cars on the road today—also contained the capacity for software upgrades to rewrite the firmware. That makes sense because, as I've stated, hardware is expensive to upgrade, so the firmware is adjusted instead. On these cars, however, there was no measure of authentication, no way for the chip to validate that the firmware being loaded was from the original manufacturer. The researchers noted in particular that it ought not be possible to rewrite the firmware on a brake component—that's a safety issue—yet they found they could. Nor should it be possible to rewrite the firmware while the car is in motion; yet, the researchers found they could do this as well.

Here we have a very sophisticated and expensive piece of machinery, one that is in virtually every driveway in the United States and indeed the world—the modern automobile—being rendered inoperable by a few random keystrokes. Even if the attacker doesn't have physical access to the target car, newer cars are using wireless vehicle-to-vehicle communication systems or vehicle ad hoc networks. This will represent yet another entry point for attackers, who could be located along the roadside.

Additionally, DARPA, an agency of the U.S. Department of Defense, is interested in fully automatic vehicles, or at least robotically controlled ones. The annual DARPA Grand Challenge is a competition to reward developers whose fully automatic or robotically controlled vehicles can endure challenges like racing through the hot desert. Such vehicles, while convenient in some situations, such as battle or the harsh environment of Mars, might also be suitable for commuters. If so, they might be a hacker's heaven, affording a variety of relatively easy attacks if the underlying chips and bus systems aren't first hardened against security breaches—that is, unless we make changes today.

6.

One of the unintended consequences of convenience is complexity. In order to make things easier, to connect to more things, we must introduce complexity. There is no easy way around it. For example, a simple system has only an on-off switch—not too convenient, is it? Think of a mobile phone with just an on-off switch. If there were no volume control, all the mobile phones today would ring in the same tone at the same decibel level, and there would be no way to silence a phone during a meeting except by powering off. Yet, by integrating granular controls such as volume, we've just made the mobile phone a lot more complex. And that's just one basic user interaction.

Complex systems are composed of individual parts, and as those parts interact, errors multiply. Only one needs to fail to permit a cybercriminal entry. As we have seen in Chapter 1, layering helps secure the system and prevent it from having a single point of failure. But are all those configurations being tested? Probably not. Companies don't have an economic incentive to expend the number of man-hours required for that level of due diligence. Nor is there a regulatory body providing technology and security standards, especially for medical gadgets. Oddly, the fault does not lie entirely with the manufacturer or with the government. Fault also lies with each of us.

In "Defeating Feature Fatigue," researchers from the University of Maryland presented participants in a controlled study with models of new audio and video players that differed only in the number of features offered. Overwhelmingly, the participants chose the most full-featured gadget as the one they would most like to own. Capability has a stronger effect on consumers than usability. A second study presented participants with a list of twenty-five features on a new audio or video player. Here the authors found that the consumer was "like the proverbial kid in a toy store," choosing more features rather than fewer. This time, however, the participants had to pay a usability penalty for each feature chosen. Even so, the participants chose 19.6 features on average. When the product arrived, however, the participants were not as satisfied with the result. "Put simply," the authors concluded, "what looks attractive in prospect does not look good in practice. Consumers often become frustrated and dissatisfied with the very cornucopia of features they originally desired and chose."[60]

Imagine a basic system with ten different settings, each with two configuration choices. When Bruce Schneier, chief technical officer at BT Counterpane, did this years ago, he found this one system alone would contain at least forty-five pairs of choices (the ways in which each configuration choice could interact with each setting) and 1,024 unique configurations overall. What if the same system had twenty settings, each with two configuration options? That's a total of 190 possible pairs of combinations and about 1 million configuration choices. And what if there were thirty settings? That's 435 combinations and almost 1 billion unique combinations overall.[61] Schneier concluded,

> I see two alternatives. The first is to recognize that the digital world will be one of ever-expanding features and options, of ever-faster product releases, of ever-increasing complexity, and of ever-decreasing security. This is the world we have today, and we can decide to embrace it knowingly.

The other choice is to slow down, to simplify, and to try to add security. Customers won't demand this—the issues are too complex for them to understand—so a consumer advocacy group is required. I can easily imagine an FDA-like organization for the Internet, but it can take a decade to approve a new prescription drug for sale, so this solution might not be economically viable.[62]

Unfortunately, there are no easy fixes here; we must take the long view. We need to start by understanding the underlying issues first. Then we must start advocating for consumer regulations that will protect us in the future. The best scenario would be for all hardware manufacturers to use authentication for software updates and encryption when communicating between subsystems. Until such a time, it's caveat emptor.

The next chapter discusses the threats we didn't see coming.

CHAPTER THREE

Invisible Threats

It's a very hot day in July 2010, and standing in front of a crowded ballroom in the Riviera Hotel and Casino in Las Vegas, Nevada, researcher Chris Paget is well aware of the illegal nature of his latest demonstration, although, with his vaguely British accent, he keeps making light of the whole situation. On stage, surrounded by a computer and electrical antennas that cost him about $1,500, Paget is attempting to emulate the signal from a Global System for Mobile (GSM) base station to attract nearby phones to his gadget instead.[1] This will allow him to eavesdrop on any mobile phone conversations taking place in the room. On the advice of lawyers, he has posted signs in the hallway informing attendees of his intentions. He has already heard from one U.S. carrier; AT&T will not sue to prevent this live demonstration at the annual security conference known as DefCon. But the U.S. Federal Communications Commission (FCC), whose representatives might be in the ballroom, has not indicated what it will do when Paget attempts to interrupt the legitimate signals from nearby towers and to record mobile phone conversations of those in the room. Depending on the FCC's definition of jamming, Paget could be breaking the law.

Paget's live demonstration was to culminate a yearlong research project. Seven months earlier, in the cold and dark week between Christmas and New Year's, Paget and his German colleague Karsten

Nohl, both from a company called H4RDW4RE, were invited by the Chaos Computer Club, one of Europe's larger hacker communities, to speak at the annual Chaos Communication Congress held at the Berliner Congress Center in Berlin, Germany. On the first day of the 26C3 conference, Nohl and Paget presented a talk titled "GSM: SRSLY?" in which they identified a number of significant flaws within the A5/1 encryption standard used for both Short Message Service (SMS) and voice messages by the most common cellular phone technology in the world, GSM.[2] They claimed that A5/1 encryption could be defeated with today's technology.[3]

A5 provides a good example of how we task old standards with new technology needs. A5/1 was first introduced in 1988 and is still in use today. Researchers first cracked A5/1 in 1998, and one year later, they cracked A5/2, a replacement algorithm. A5/3, which is entirely different from both A5/1 and A5/2, has yet to be deployed worldwide.[4] Nohl and Paget decided to rally against the GSM industry, which continues to use the weaker, and academically broken, A5s.

It must be noted that the breaks within A5 don't necessarily apply to how GSM cellular networks implement the encryption. New flavors of third-generation (3G) GSM, such as General Packet Radio Services (GPRS) and Enhanced Data Rates for GSM Evolution (EDGE), for instance, have made communications a bit more secure, but Paget still maintains there is "not even a pretense of secrecy in GSM."[5] For example, A5/1 is not as secure as other forms of encryption, such as the Data Encryption Standard (DES) and Advanced Encryption Standard (AES), both of which have since been replaced with Triple DES and AES 256 today.

Now that we're asking our mobile phones to handle banking and some e-commerce, shouldn't we make sure the platforms are secure first? Standards should be strengthened to accommodate our growing need for mobile commerce. Even the Internet (currently based on a standard known as Internet Protocol version 4 ratified in 1981) was not designed for e-commerce; yet, we buy and sell on the Internet today.[6] As we saw in the previous chapters, networks can be segmented but are more often bundled together, such as those we saw in hotel rooms, becoming more complex as needs evolve. This is no less true with the cellular networks we have today.

Perhaps you've seen ads for competing cellular services stating they have the fastest 3G network or introducing fourth-generation, or 4G, networks. What does that mean? First generation, or 1G, refers to the technology available circa the 1980s; these early networks and handsets were analog and used a variety of mobile standards, some no longer in use today. In 1991, the second-generation (2G) network was introduced, a digital network with two standards, GSM and Qualcomm's Code Division Multiple Access (CDMA). Both GSM and CDMA include SMS, Unstructured Supplementary Services Data, and other simple communications protocols.

Of the two standards, CDMA is considerably harder to eavesdrop on.[7] CDMA doesn't send its radio signals intact but divides the signal over a range of channels. The receiver then takes these individual packets and reconstructs the message. Originally designed for military use, CDMA provides a secure communications system that separates users by generating random codes that are never transmitted. Adding encryption makes intercepting a CDMA call virtually impossible. Dominant in North America, CDMA is used by Sprint and Verizon in the United States.

GSM, on the other hand, is used by 80 percent of the world's mobile phones. In the United States, T-Mobile and AT&T use GSM. So, why are most of the GSM networks in the world still using the weak and vulnerable A5/1 when A5/3 is much harder to break?

First, said Nohl, not every country uses encryption; for example, India and China do not encrypt their GSM networks. Second, the GSM Association has found that upgrading all the cellular towers and network equipment to accommodate the additional resources needed for the additional encryption isn't always feasible. For example, in 2004, when a critical vulnerability was found in A5/2, cellular operators in South America, Africa, and parts of Asia required about eighteen months to upgrade to the more secure equipment.[8] Upgrading everyone around the world to support the stronger A5/3 standard will not be any easier.

In the summer of 2009, Nohl called on fellow researchers to help generate all the possible A5/1 encryption codes used by the phone carriers.[9] Up until now researchers could not replicate all the computation cycles used by the phone carriers to compute all the different variations of the encryption keys for A5/1. By using a distributed table generation method, Nohl and Paget got individuals all over the world to contribute minutes of spare computer time, and within six months, by December 2009, the two had assembled a comprehensive lookup table of all the possible values for A5/1 encryption in the GSM network.[10] The resulting table, known as a rainbow table, is large—over 2 TB in size. This is unwieldy for most. Nor does such a large table allow for near-instantaneous lookup.

Nohl then spent the next few months working on ways to make his lookup table more efficient and found a method that allowed him to store only the input and the output, chaining all the possible combinations elsewhere, resulting in a much lighter and faster rainbow table, which he debuted at Black Hat USA in July 2010.[11] In his demonstration, Nohl showed an encrypted string of numbers captured from his own mobile phone conversation on its way to a base station he controlled in the lab. He then, in a matter of a minute or so, decrypted the call and played his voice for the live audience.

A few days later, Paget's DefCon demo would take things even further. Paget planned to stream and decrypt a live conversation captured over the GSM network. How would that work?

Mobile phones use radio waves to transmit and receive data such as telephone conversations and text messages. Many of the same principles that affect traditional radios also apply to Wi-Fi and mobile phones. For example, anyone can transmit on a particular channel so that it's difficult for a receiver to know if a given transmission is legitimate or spoofed.

Before Nohl and Paget each spoke at Black Hat and DefCon, another researcher, known only as "the Grugq," showed how to spoof signals to a legitimate GSM base station using only an inexpensive and reprogrammed GSM mobile handset.

Whenever a mobile phone call is made, the Grugq said, the handset establishes a handshake with the nearest cell tower, except 2G GSM phone networks lack mutual authentication.[12] That means a mobile handset authenticates itself to the network (it says, "I have a chip that allows me access onto your network"), but the network doesn't have to authenticate to the mobile handset (it doesn't say, "Yes, I am the proper network for your handset"). The mobile handset must provide a secret key to the cellular tower to gain permission to talk to the network, but the network doesn't have to prove its legitimacy. That lack of mutual authentication creates some interesting opportunities.

The Grugq showed how he could use his inexpensive GSM-enabled phone to wreak all sorts of havoc on nearby GSM base stations.[13] For example, if one knew the unique identifier assigned to a particular phone, that is, the unique code assigned to each mobile phone, one could tell the network that the phone's owner had disconnected, then repeatedly send the disconnect signal to keep that person off the network entirely.

Phones send out a signal called an International Mobile Subscriber Identity (IMSI, pronounced "IM-SEE"). The IMSI lives in the subscriber identity module, or SIM, card, which is the chip moved from handset to handset when you upgrade your phone. The IMSI is broadcast to the network every so many seconds to let the network know where you are and whether you are still powered up. The attack the Grugq outlined sends an IMSI "detach" command for your specific phone. The network, seeing that you have detached, sends all incoming calls to voicemail, so no incoming calls are lost; you just won't be able to make any outgoing calls resulting in a denial-of-service attack against one phone.

Conversely, you could use the Grugq's attack to send a request to connect, an IMSI "attach" command, with no follow-up handshake. This floods the base station or cell tower with requests to hold open a channel while waiting for a response that will never come. By exhausting all the open channels, you prevent anyone from connecting to the base station, creating a denial-of-service attack against not only one phone but all phones in the area. (The Grugq said he did this by accident at work, and the company was without cellular service for a day.[14])

The police in Europe use a device that works similarly to this called an IMSI catcher. The intent is to capture IMSIs, not deny service. An IMSI catcher is a form of base station. Because the IMSI catcher signal

is stronger than the nearest tower, mobile phones will attempt to connect through the IMSI catcher first. Once the IMSI catcher records the
initial request, it lets go of the call so the handset is free to connect to
the real base station. Using the captured IMSIs, the police then work
with the carriers to identify those who attended the rally.

However, an IMSI flood attack would thwart IMSI catches by flood-
ing the IMSI catcher with bogus call attempts.

At DefCon, Paget wanted his base station to complete the mobile
phone handshake so that he could intercept a live call. He could have
further connected his base station to a live network base station and
performed what's called a man-in-the-middle (MitM) attack, forcing all
calls to be passed through and recorded by his machine, but he chose
not to do so for this demonstration.

To circumvent some legal issues, Paget broadcast his GSM base station at 900 MHz. This happens to be a ham radio frequency, and Paget
happens to hold a ham radio license. In the DefCon demo he told the
base station to emulate AT&T's signal, and about two dozen phones
then connected to Paget's system instead of the nearest legitimate
AT&T tower in Las Vegas.[15] During the demo, Paget broadcast a message telling the cellular customer his or her call would be recorded.
After the talk, Paget destroyed the USB stick containing all the recordings with a pair of scissors.[16]

Paget's demonstration only works on 2G networks. To lure 3G-
enabled GSM phones, however, he simply jams the 3G signal, which
forces the handset to attempt to connect with a 2G signal. Here, legacy
systems play into criminals' hands. Paget's advice: Turn off 2G on 3G
phones, if possible. But that would also mean loss of service in some
non-3G areas.

When connecting a handset to a cellular network, the base station
currently chooses what level of encryption will be used. In his demonstration, Paget simply turned A5 encryption off all together. GSM networks are supposed to inform you when encryption is not in use, he
said, adding that most carriers turn this message off.

After Paget's talk, the GSM Association told Forbes.com, "The overall advice for GSM calls and fixed line calls is the same. Neither has ever

offered a guarantee of secure communications. The great majority of users will make calls with no reason to fear that anyone might be listening. However, users with especially high security requirements should consider adding extra, end-to-end security features over the top of both their fixed line calls and their mobile calls."[17]

If mobile phone conversations can be intercepted without our knowing, what are other gadgets capable of, and what other personal information might we be leaking?

1.

Maybe someone won't be eavesdropping on your GSM SMS or voice messages, or perhaps you use a CDMA provider like Verizon or Sprint for your mobile service. There is, however, a much more common concern. An ad for real-time banking once showed two guys sitting in an Internet café. One was online checking his bank account. The other was using his friend's debit card to buy a latte. The point of the commercial was that as soon as the debit card had been swiped at the cash register, the card owner could see the transaction go through online. The problem I and others had with the commercial, however, was that one of the nation's largest banks was advocating that people should check their bank account balances while using a public Wi-Fi network.

A cybercriminal can eavesdrop on a public wireless session by performing a man-in-the-middle attack. This requires setting up a duplicate public access point (AP) using another computer (typically a laptop, although such an attack might also be possible with a smart phone).[18] The attacker adds an antenna to the laptop, then rebroadcasts the actual settings of the legitimate AP so that an unsuspecting café or airport patron logs on to the stronger fraudulent signal—this is much like what Paget was doing with his fake GSM base station, acting as a man in the middle of a cellular transmission. Unless your firewall informs you, you won't necessarily know this has happened with a Wi-Fi connection. While you do connect to the Internet, you do so through the cybercriminal's laptop. To the end user connecting through a criminal's laptop, the Internet experience is no different from normal.

Also dubbed "evil-twin attacks," these MitM attacks allow a cyber-criminal to "sniff," or read, any data that the victim is sending via the Internet, such as the login ID and password for an online banking account or any e-mail sent without encryption.[19] A few years ago the end user could protect him- or herself by switching to a virtual private network (VPN), which encrypts the connection, denying the attacker the ability to eavesdrop. Today, criminal attacks are getting more and more sophisticated; cybercriminals can grab the data before encryption in what's called a man-in-the-browser attack by installing malware onto your computer.

The convenience of public Wi-Fi is possible, in part, because of arguably dangerous defaults. If you're just surfing the Web, looking for sports scores or the weather in a foreign city, then you aren't risking too much on a public Wi-Fi access point. But if you log on from a free, public Internet café, hotel, or airport waiting area to buy a present for your kid or check your stock portfolio, then you're risking a lot. Unlike wired Internet connections, wireless signals can be captured or sniffed by others with the right equipment. (Wired signals can also be sniffed or captured, but that requires an attacker to gain access to the wired local area network first.) New technology uses multiple means of ensuring a constant Internet connection. This means your mobile handset is automatically switching between different "radios" within your mobile phone handset to stay connected. Some phones "bridge" between two networks.

Let's start with a basic wireless Internet public system. First, access must be convenient, so the identity of the AP must be clear, broadcasting "Airport Café" for all to see. Sometimes there is no password to make it easier for random customers to use the network when in range. And when there is a password, often it is obvious or easy to remember, such as "123MainStreet." Unfortunately, Windows remembers these commonly used APs in order to link with these Internet sources later when you want to reconnect. This is convenient if you've been on a business trip and want your laptop to work instantly with your home network. But there's an obvious dark side to such convenience.

Even if you don't use an Internet café, most people never change their home router's default information, so there are plenty of "Linksys" and "Netgear" service set identifiers in the world. A criminal could broadcast one of these common router names and connect to your wireless laptop. Microsoft patched this flaw, but while the software company turned off the external broadcast list, Windows XP systems still connect to an ad hoc network if its name agrees with the internal list kept by the laptop.[20]

How might this be prevented? Another unique identifier assigned to each gadget is the media access control (MAC) address composed of six octets, which contain eight bits each.[21] The first three octets usually identify the manufacturer, while the remaining three are specific to the gadget itself.[22] So, a gadget can additionally be fingerprinted using its MAC address. Thus, a router can be configured to connect only with specific MAC addresses; this works for a home router but not a public router, to which a great many unknown gadgets will likely want to connect.

Connecting to a wireless network is complex, and router companies would not have much of a market if they did not create shortcuts and simplify the process. In 2006, a report commissioned by the Consumer Electronics Association found that almost one in ten customers had returned a home network router, hub, bridge, or modem within the previous year. Of those returned gadgets, only 15 percent were truly defective; the other 85 percent had been returned because the average consumer simply could not figure out how to make them work.[23]

So what did the router companies do? They created wizards that walk users through the basic steps necessary to connect to their Internet service providers. Along the way, these wizards don't mention all the possible security configurations; manufacturers just want consumers to connect—and stop calling their customer-service lines or returning the gadgets. While these home networks sped up Internet adoption and perhaps hastened the popularity of Web 2.0 applications like Gmail and Facebook, a number of gadgets connected to them without additional Internet security software installed were left vulnerable to malware.

Let's go back to our MitM public Wi-Fi scenario. Pretend I'm interested in some data that your company is working on (a financial report, a customer list, or a prototype of a new product). Through a site such as LinkedIn, which professionals use to post their resumes, I can learn your name and position. I can then go to a site such as Foursquare, which broadcasts your current geolocation and over time tracks your habits. I can, for example, learn that on Wednesday you "work from home" and glean which Starbucks wireless location you use to log in to the office. I can arrange to be at that Starbucks on Wednesdays, and maybe I can also defeat its wireless network so that your laptop thinks my laptop is a legitimate connection. This requires a lot of work perhaps, but we've already seen the lengths to which cybercriminals will go to attack someone.

Now, let's apply this to smart phones. Internet access via 3G networks provides smart phones with faster data flows, ushering in a new group of standards, all designed around improving data transport and keeping the end user's connection with the Internet. With 3G, these standards include Evolution–Data Optimized, also known as CDMA2000, and EDGE. Additionally, for GSM, there's the GPRS standard, which is designed to complement existing services such as circuit-switched cellular phone connections and SMS. GPRS can bridge different networks (for example, 2.5G and 3G). Packet-based systems bundle data into groups that can be queued. Circuit-based systems open a channel. Packet-based systems cost less to operate since the packets are allocated as needed and can be shared with others, whereas in a circuit-based system, one user ties up an entire channel for an indefinite period.[24] GPRS offers data rates from 56 to 114 Kbps and continuous connection to the Internet, which is helpful for interacting with multimedia websites or streaming interactive videoconferences on the mobile gadget.

By default, many mobile gadgets today support more than radio systems, among them GPRS/EDGE and 802.11 (aka Wi-Fi); this is so that end users receive uninterrupted data and Internet service no matter where they roam. One of the first signs of a possible mobile MitM attack, said Paul Henry, a security and forensic analyst at Lumension, is a pop-up certificate for a website or portal that you regularly visit. "That you are prompted to accept a new certificate on a familiar site should be a red flag," he said.[25]

Because we're using a mobile phone, Henry said, we simply accept the new certificate and carry on with our business, unwittingly allowing messages to be intercepted from that point on. Wi-Fi-enabled smart phones create some interesting attack vectors. A smart phone connected via Wi-Fi might be subjected to the MitM attacks discussed above. If you're conducting mobile banking, having someone eavesdrop on an unencrypted banking session could be a real problem. Even a man-in-the-middle attack within a GPRS network is a credible real risk, Henry said. In another scenario, he added, someone could be connected via Wi-Fi to a local AP inside a corporate campus and also use the GPRS connection to the outside world, effectively mining the corporate network from afar.[26]

2.

Any Hollywood heist film almost always includes a scene in which the criminals patch a static video into the surveillance camera network showing an empty bank vault full of money. This video masks the criminals as they dash inside and completely loot the vault (see, for example, Steven Soderbergh's 2001 remake of *Ocean's Eleven*). Of course, someone needs to be on the inside of the organization or at least physically on the premises to do this. In 2009, Jason Ostrom, director of Sipera Viper Lab, demonstrated how this could be done remotely with Video over IP systems.[27]

Like any other technology, Video over IP must be configured securely, a process called *hardening* within the security community. Yet, many systems are installed with default settings intact. Ostrom, who works as a security penetration tester, has found that only one in twenty organizations he's seen has encryption on its video systems. So, he and others at Viper Lab created a tool, UCSniff, which looks for the presence of a video stream and evaluates its security, while another tool, VideoJak, injects video, such as a static shot of the full vault, into a video stream.[28] These tools, created using open-source programs readily available on the Internet, were designed to show organizations the consequences of poorly configured systems.

While Ostrom's work focuses on the enterprise system, Adrian Pastor's research in the United Kingdom has focused on public street-corner surveillance systems used in that country.[29] Using this combined work, attackers could bypass both a street-corner security camera and internal security cameras so that local law enforcement wouldn't see them coming and going or what they did while inside. Remarkably, both sets of attacks could be pulled off remotely by someone not physically near the building being robbed.

Pastor and Ostrom also overlap on Voice over IP (VoIP) research as well.[30] VoIP is popular because over the Internet there are no toll calls, and less cabling is involved, so employees can switch offices without the telecom people having to rewire or even change switching operations on their phones. Finally, because the system is digital, integration with e-mail clients is possible, and voice mail notices can appear in one's Outlook inbox.

Lumension's Paul Henry has seen default VoIP installations in Southeast Asia that pay no regard to the risks. Management sees only the potential savings. He cited one positive example: When moving into its new office building, Thailand's Securities and Exchange Commission (SEC) built out its network infrastructure with care.

"The head of the SEC," Henry said, "was very interested in how easy it was to capture a VoIP voice or video call and commented that VoIP can facilitate insider trading." This could have serious consequences for the Thai stock exchange. The head of the SEC commented "that in their environment, VoIP can potentially hinder their investigations if the calls are captured between their teams of investigators," said Henry.[31] Unfortunately, not every organization has performed such due diligence.

Ostrom has demonstrated just how VoIP can be harnessed to expose a business's internal network to outsiders.[32] Here, an attacker would need physical access to a phone hooked up to the network—one, for instance, in a public waiting area, conference room, or hotel lobby. He would then have only to unplug the house phone and plug in his laptop instead.

In previous work done with John Kindervag, senior security architect for Vigilar, Ostrom created a tool known as VoIP Hopper to show com-

panies just how vulnerable they are. In particular, VoIP Hopper inter-
cepts calls made with the Cisco Discovery Protocol (CDP). When pen-
etrated by VoIP Hopper, CDP will also create a new Ethernet gadget
that could allow someone to map or otherwise damage a company's
network from a public waiting area.[33]

The VoIP Hopper tool allows you to remove the phone physically
and use a laptop to spoof the phone's MAC address. In other words,
you make the network think the laptop is just another phone. This is
important: Just because the network sees a telephone MAC address,
that doesn't mean a telephone is attached to the network. This is similar to identity fraud in which someone sends you an e-mail that looks
like it's from a person you know, perhaps right down to using your
friend's actual e-mail address. The VoIP network is unaware that a laptop has replaced the phone; thus, the laptop can engage in a MitM attack, listening in on all conversations.

How are Ostrom's and Pastor's attacks possible? Address Resolu-
tion Protocol (ARP) is used when gadget A on a network needs to send
a job to gadget B. For example, desktop A sends a print job to printer B.
ARP combines a device's MAC address with its IP address so that all
gadgets on a network can find each other. But there is no authentication. Thus, one gadget can pretend to be another. Known as ARP poisoning, this technique has been around for years. ARP poisoning is one
way a laptop can masquerade as a public telephone on a VoIP system.
To prevent such attacks, Ostrom recommends turning off unused features and disabling unnecessary ports on public VoIP phones.

3.

The previous chapter discussed how design flaws in hotel networks allow
tech-savvy guests to use the infrared-based TV remote to eavesdrop on
the hotel billing system. In 2007, San Francisco–based researcher Luis
Miras toured the security conference circuit with his own analysis of
vulnerabilities in other common radio-based gadgets, such as wireless
keyboards, mice, and presentation pointers.[34] For example, he showed
how someone could disrupt a PowerPoint presentation.

Like the basic hotel TV remote, the radio frequency (RF) gadgets Miras studied all enjoyed "security by obscurity"; that is, no one's going to bother to hack it because no one thinks about these systems. Miras found that many of these convenient radio frequency–based computer gadgets don't follow accepted protocols and standards; therefore, they lend themselves to interception by others.

The wireless keyboards, mice, and presenters, when broken down, contain only a few components. Take, for example, a wireless pointer, which will contain only a microcontroller, an electrically erasable programmable read-only memory (EEPROM) chip, a transmitter chip, and some means of input (e.g., a button, mouse movements). The EEPROM chip holds the identification number. These gadgets often use less expensive chip sets and perform only one-way communication: They don't accept commands from a PC; they only send them. But there are other ways to attack these gadgets.

The "brains" of the ergonomic wireless mouse are inside a tiny microcontroller that takes in user input (left, right, up, down, click), then sends that data to the transmitter, which in turn simply sends whatever data it is given as a radio signal.

On the PC, the companion USB receiver contains a chip that converts the RF signal back into bits. The USB's microcontroller then reads the gadget's identification number to determine whether the data is intended for its use. If the identification numbers match, the USB passes the data to the computer.

A computer operating system groups these wireless gadgets (keyboards, mice, and presenters) under the type human interface design (HID). The design flaw is that default device types are used indiscriminately. For example, a wireless presenter, used to advance slides in a PowerPoint presentation, often registers itself as an HID keyboard. In order to reduce costs further, the full command for "page up" and "page down" is not presented; rather, a proprietary command is used instead. Basically, the "forward slide" and "backward slide" functions are presented to the computer as "page up" and "page down" commands.

Mice present a different situation, said Miras. A specific HID device type for mice allows for absolute movements, such as moving the mouse diagonally or drawing. Keyboards use the keyboard device type. The key commands on a keyboard are converted into a proprietary command structure sent to the computer.

These are all gadgets that broadcast, and every broadcasting gadget sold in the United States receives an FCC number, which can be reviewed on the FCC's website.[35] There the FCC requires manufacturers to provide schematics, external and internal photographs, a user's manual, and other technical data. While the manufacturer can demand confidentiality, that rarely happens. Often there is a treasure trove of technical information on the FCC site for those who seek it.[36]

Miras said that there are two types of attacks against these gadgets. One requires listening to, or sniffing, the RF signals broadcast between the gadget and the computer. By learning that a Kensington Mouse always uses a particular sequence of ones and zeros for an action, one can "eavesdrop" on where the mouse is oriented on a given screen. Far more practical would be eavesdropping on a wireless keyboard. Here passwords and other sensitive information can potentially be mapped back to the original keys pressed.

The second type of attack, an active attack, is arguably more useful. Here the communication between the microcontroller and the transmitter is hijacked, and a third party is able to inject commands directly into the receiver: a classic MitM attack. One result could be a denial-of-service attack—simply rendering the mouse, keyboard, or presenter useless. The attacker could also take over the receiver remotely to execute his or her own commands. Think of a PowerPoint presentation with a pointer gone awry, the speaker, frustrated, having to shut down the computer and talk extemporaneously—a lost skill.

But these attacks are hard to pull off. Most microcontrollers are one-time programmable, meaning an outside source cannot overwrite the programming burned onto the chips. Also, these gadgets have limited uses; they must be operated within a line of sight. The infrared signal cannot penetrate walls; thus, the attacker must be in the same room and in close proximity.

Some attacks can be run blind, meaning that the criminal need not be able to see the display screen but can rely on replay attacks (where a legitimate session is recorded, then later replayed). This works with simple systems such as garage door openers.

Having your PowerPoint presentation hijacked might not seem like much of a risk, but using the FCC's information, Miras found that in addition to mice, keyboards, and presenters, certain car alarms and wireless home security systems that use many of these same radio-based components are vulnerable. With a little work, attacks against these systems might also be possible. Suddenly radio signals are not so trivial.

4.

Sometimes eavesdropping on gadgets is unintentional, a natural byproduct of rapid innovation and adoption. When cordless phones first appeared on the U.S. market around 1980, they were too popular for the narrow frequency assigned by the FCC.[37] The earliest cordless phones had default frequencies at 27 MHz on the low end of the radio spectrum. This created several problems, partially the result of having to use analog phone signals. While cordless phones allowed people to talk, their range was limited—you couldn't go too far from the base unit. And the sound was poor because the signal had to pass through walls and other appliances in the home or office. It was essentially a party line with multiple people attempting to make a call at the same time in the same proximity such that they could all hear each other on the limited channels allotted to those early phones.

By the mid-1980s, the FCC granted cordless phones a new frequency range of 47 to 49 MHz. While this helped somewhat with the interference problem, it still left much to be desired regarding sound quality and range. By 1990, the FCC had opened the 900 MHz frequency range.

This had a few significant effects. It improved the overall range of the phones and also opened them to multiple channels within the frequency, so two phones could share the same frequency but different channels. Thus, no more party lines.

By the mid-1990s, cordless phones had gone digital. Introduction of the direct-sequence spread spectrum (DSSS) technology made eavesdropping much tougher. DSSS allowed the voice signal to be spread over several channels within a frequency, making it harder for someone to reconstruct the signal in order to eavesdrop. A gadget chooses a random channel with DSSS and then flips through a spectrum of channels within the frequency every second. If every gadget chooses a different channel, and if every gadget flips through the spectrum of channels every second, there should be no interference. Thus, two different 900 MHz phones could operate side by side. By the turn of the twenty-first century, the FCC had opened up the 2.4 GHz range.

Early cordless phones using the low-end 47 to 49 MHz frequency once cost about $400. The cost of those transmitters has come down sharply, and now they are to be found in $50 baby monitors sold at Babies"R"Us. If we had problems with interference and eavesdropping on these phones back in the 1980s, it seems likely the same condition will persist with the baby monitors today.

In late 2009, Chicago resident Wes Denkov filed a class-action suit against the manufacturers of the Summer Infant Day & Night Baby Video Monitor.[38] In the suit, Denkov said he had bought the monitor for his infant son's nursery. Then, about six months later, his neighbors, parents of newborn twins, noticed that they could see and hear the video and audio from the Denkovs' monitor. The suit claims that when the neighbors turned the channels (there are two channels on the product), they could sometimes observe all the conversation and activity within the Denkovs' nursery.

This isn't too surprising since the 47 to 49 MHz range is also accessible on most police scanners available at the electronics store, and a police scanner within range can also eavesdrop on a baby scanner. Here we have an example of a known default setting, the limited ability of the transmitters to offer security, being pushed down to an inexpensive baby monitor. But the use of older, outdated technology in pricier mainstream gadgets is also quite common.

Crowded frequencies can have life-and-death consequences. During World War II, U.S.-made radio-guided torpedoes were subject to frequency jamming by the Germans. One solution, known as frequency-hopping spread spectrum (FHSS), was actually developed and patented by an actress, Hedy Lamarr. Lamarr was married to an arms merchant and as a result attended several defense contractor meetings. One day she and her husband, George Antheil, realized that to prevent jamming, one only had to change the frequency with each use; so long as the boat and the torpedo were always on the same frequency, the signal could not be jammed (both the torpedo and the boat would have copies of the same frequency order). In 1941, Lamarr and her husband received a patent for their idea.[39]

FHSS was later incorporated into CDMA, a frequency-hopping technique used for radio communication. This is slightly different from the Qualcomm standard used for mobile networks mentioned earlier in this chapter, although it was used with cdmaOne, Qualcomm's first CDMA standard, and in the current CDMA2000 network standard for mobile phone channel allocation.

Another solution that addresses overcrowded frequencies is Bluetooth technology. Bluetooth, named after the tenth-century Danish king Harold Bluetooth, is omnidirectional and governed by the Bluetooth Special Interest Group, a consortium of companies.[40] Bluetooth also uses the 2.4 GHz frequency range and has a limited range of about ten meters (thirty-two feet).[41] The Bluetooth range can be extended— and extended by a lot with additional hardware.

Bluetooth in particular employs adaptive frequency-hopping (AFH) spread spectrum, which uses seventy-nine distinct, randomly chosen frequencies and changes frequency 1,600 times a second.[42] To make AFH, Bluetooth generates a pseudorandom number based on what is called the Bluetooth device address (BD_ADDR), which is similar to a MAC address.

Bluetooth can connect up to eight gadgets at once, forming what is called a personal area network or piconet. Within each piconet there is one master. All Bluetooth gadgets can be either a master or a slave. The AFHs for the individual slaves are derived from the one master BD_ADDR. Similarly, a slave in one piconet can be a master in another. A group of two or more piconets is called a scatternet. Thus, invisible webs of Bluetooth gadgets can exist within a single room.

Knowing the master's BD_ADDR is essential for a slave to derive its AFH pattern. Bluetooth might have had an HID so that users could enter in the information—but that would not be practical; people aren't going to type in forty-eight characters. Pairing two gadgets needed to be automatic.

Instead, Bluetooth was designed to probe another gadget's BD_ADDR in what's called discoverable mode. When two gadgets pair, they exchange a PIN. This PIN, along with the BD_ADDR, is then used to create a 128-bit encryption key. A link key is then shared on both gadgets and used each time the two gadgets want to communicate. Thus, the PIN is encrypted only at authentication or pairing and not throughout each transmission.[43]

Bluetooth relies on the secrecy of the BD_ADDR, although software exists that can scan for BD_ADDRs in the local area. With one of the programs, potential victims receive an SMS asking them to turn off their Bluetooth and restart. When they do, the gadget is briefly vulnerable as the BD_ADDR is exposed.

In the early days of Bluetooth, PDAs and then mobile phones had Bluetooth discoverable mode enabled by default. There were advantages to this. The limited number of Bluetooth-enabled gadgets made it possible and convenient for people to connect with each other. One feature, Bluetooth Object Exchange (OBEX), allows for the convenient exchange of business cards wirelessly at a conference. It is similar to Hypertext Transfer Protocol (HTTP), used for the Internet, in that it allows a gadget to request objects. The OBEX push profile, however, doesn't allow for authentication.

Not surprisingly, there are significant problems with this feature.

One problem is Bluejacking, an easy prank to pull off. Simply create a new phone contact in your mobile gadget phonebook and, instead of putting in a name, type a short message, such as "You've been pwned!" Now, scan the local area for a Bluetooth mobile phone that's in discoverable mode and send the contact to them. Instead of "Bob is calling you," the owner of that phone will see the message that he or she has been owned by a cybercriminal. But Bluejacking is harmless. It basically uses Bluetooth to send SMS-like text messages. The first known use of a Bluejacking came from a Malaysian IT consultant who broadcast the words "Sony Ericsson" on a Bluetooth-enabled phone as a marketing ploy.[44] In Europe, home to Nokia and Sony Ericsson and where Bluetooth phones were first available, Bluejacking reached cult status and was used to initiate sex by sending explicit messages.

A more pernicious Bluetooth attack is called Bluesnarfing. While studying the security of Bluetooth, Adam Laurie, whose hacking of the hotel TV remote was described in Chapter 2, found a way for one Bluetooth gadget to collect files from another Bluetooth gadget. The attack was independently discovered by Austrian researchers Martin Herfurt and Marcel Holtmann. Holtmann is also a member of the Bluetooth Special Interest Group's Security Experts Group.[45]

To Bluesnarf, an attacker needs a connection to OBEX Push Profile, which allows for phone-to-phone transfers, and a program called Bloover. (Get it? Bluetooth combined with Hoover because the program sucks up the files on another gadget.) Bloover was written by Herfurt. An attacker performs an OBEX "get" request for a known filename, for example, the mobile gadget's phone book. Ordinarily, this would be a call for a file named "[telecom]/pb.vcs." Bluetooth would respond by sending that file to the requester. On some phones it is possible to request all the files.

Bluesnarfing is illegal in many countries. Also, many mobile gadget manufacturers have changed the default Bluetooth setting from discoverable to hidden unless the user requests otherwise. Moreover, the Bluetooth standard has since been updated such that gadgets must first be paired, that is, authenticated with each other, before an OBEX file transfer can occur. While Bluetooth gadgets are set to nondiscoverable mode by default, during a reboot, a Bluetooth gadget is briefly in discoverable mode and thus vulnerable to attack. Ironically, most Bluetooth passwords for embedded devices such as headsets are also set by default to either "0000" or "1234" for convenience.

Previously, I stated that Bluetooth's range was limited to ten meters but could be extended. The current record is over one hundred miles.[46] From a mountain near Las Vegas, a research team was able to contact a Bluetooth gadget over 155 miles away near St. George, Utah, using two satellite dish antennas—one measured twelve feet, the other ten feet—placed on either side of a Bluetooth PCMCIA card in a laptop running Linux. Previously, a team including Martin Herfurt, Mike Outmesguine, John Hering, James Burgess, and Kevin Mahaffey held a record of 1.08 miles in Santa Monica, California, using a directional antenna mounted in a sniper rifle configuration, which included, according to Mahaffey, a 24 dBi antenna with a modified class 1 dongle.[47] In the process the team could have connected to other Bluetooth-enabled phones, but they attacked only the target phone they'd planted themselves.[48]

So, what about other Bluetooth-enabled gadgets? Could your infected Bluetooth phone infect your Bluetooth printer or DVD player? Back in January 2005, antivirus researcher Eugene Kaspersky said that his company had been asked to investigate whether cell phone viruses could infect Bluetooth-enabled Lexuses in Russia. "If infected mobile devices are scary, just think about an infected onboard computer," he's quoted as saying.[49]

Increasingly, today's cars are more computer based than mechanical. Toyota, Ford, and other auto manufacturers employ Bluetooth technology to transfer phone book and music information from your mobile phone to the car's own speaker system, allowing for hands-free conversation and music selection. We saw in the previous chapter how Ford is securing its Bluetooth technology in its cars.

Despite the front-page media buzz, Kaspersky produced few facts. The story went cold. However, Mikko Hypponen, chief research officer at F-Secure in Finland, remained curious. Could a mobile virus leap from one Bluetooth gadget to another? He decided to investigate. Unable to obtain a Lexus, Hypponen settled on a 2005 Prius from Toyota (the company that also owns Lexus). The Prius's Bluetooth system was a proprietary operating system different from the mobile phone's. To eliminate ambient Bluetooth broadcasts in the area, Hypponen and his crew from F-Secure Labs tested the Prius in a bunker deep within the earth just outside of Helsinki.

Despite various scenarios in which someone enters a Bluetooth-enabled Prius with a malware-infected mobile phone, the team at

F-Secure was unable to infect the Prius's Bluetooth. And when one of the researchers attempted to send an infected file to the car, the Prius responded with a "transfer failed" message. In fact, they were unable to transfer any known Bluetooth attack from one gadget to another.

At worst, the car's entertainment system computer simply rebooted.

5.

There is, however, one known malicious connection between a Bluetooth mobile phone and a Bluetooth-enabled car. With the right software, such as Martin Herfurt's Car Whisperer, you can hear conversations held in the Bluetooth-enabled car ahead of you when the driver's mobile phone is not in operation. In other words, the Car Whisperer can eavesdrop on conversations. You only need a Bluetooth USB, an antenna, and a Linux computer. The Car Whisperer can also inject audio into a car's speaker system, such as "Attention! This is the police. Please pull over now and come out with your hands up."

The problem isn't with Bluetooth itself but rather with its implementation in the cars. Herfurt observed that most auto Bluetooth systems use default PINs such as "0000" and "1234" to make pairing with mobile gadgets as convenient for the end user as possible. This is similar to the wizards used in home network routers mentioned earlier; if pairing Bluetooth were hard, people wouldn't do it. Thus, Bluetooth pairing with a car can be as easy as guessing the four digits issued.

The Car Whisperer works by first scanning for the presence of Bluetooth gadgets. It then looks for gadgets of a certain class or type. By looking at the first three digits of the MAC address returned, the program can determine the particular manufacturer. It then knows what PIN to try to pair. Once paired, the Car Whisperer connects to the car's Bluetooth system.

The beauty of this attack is that it doesn't attempt to dislodge the mobile gadget already in use. Rather the Car Whisperer adds a gadget. Remember, a Bluetooth device can support up to eight different connections. The Car Whisperer sends the target Bluetooth car kit a message saying, in effect, that the Linux laptop has lost its key and asking the car's Bluetooth kit to please send the key again. Without authentication, the Bluetooth system has no idea whether the laptop is making a legitimate request. Once the Car Whisperer receives the key, the attacker is able to pair as yet another Bluetooth gadget.

This attack can allow law enforcement to listen to someone in the Bluetooth-enabled car ahead when his or her mobile phone isn't engaged in a call. Or, as noted, it can allow a custom message to be played over the car's speakers. The Car Whisperer, available as open-source software on the Internet, can also allow anyone with the technical chops to listen as well—it's currently legal.

The Car Whisperer has other uses. Researchers have realized that since it doesn't dislodge existing gadgets, you can use the Car Whisperer program at Starbucks to pair a Linux laptop with a Bluetooth hands-free headset. As with the car ahead, the Car Whisperer can scan for the presence of a Bluetooth gadget. As long as the party isn't on an active call, the laptop can pair with the headset, and through its microphone, the attacker can overhear, for example, someone making small talk with the baristas or, more ominously, a business transaction going on two tables down.[50]

Sometimes technologically enhanced eavesdropping is far more serious than mere espionage.

Since 2002, the U.S. military has been inadvertently sharing highly sensitive video with the enemy in Afghanistan, Pakistan, and Iraq. Shortly after the start of the wars in Afghanistan and Iraq, the U.S. military started flying Predator drones and AC-130 gunships over the mountainous countries. The Predator drones in particular allow for beyond-visual-line-of-sight reconnaissance, surveillance, and target acquisition—important information for troops on the ground. Taking a shortcut in the interests of getting the video to the ground troops quickly, the military developed the Remotely Operated Video-Enhanced Receiver (ROVER) in 2002. This shoulder-carried gadget receives the satellite-band channels used by the unmanned and later manned aircraft without any encryption.

A November 2008 U.S. military presentation stated that the ROVER 4 signals were unencrypted and vulnerable to eavesdropping by the insurgents.[51] According to Ron Smith, chief of the Logistics Division, Unmanned Aircraft, Project Managers Office, enemy forces have been able to view the satellite images and ambush American troops for years using a satellite dish along with a $26 Russian-made software program called SkyGrabber.

Since 2001 the United States has fully adopted ROVER and turned virtually every aircraft in Afghanistan, Pakistan, and Iraq into "an eye in the sky" using this unencrypted signal. Nearly every commander on the ground has become accustomed to the technology. As mentioned previously, switching over hardware in the face of a vulnerability disclosure is sometimes not feasible. Adding encryption to ROVER will introduce a slight time delay, which in a war theater could also be deadly. But pretending a flaw as dangerous as this one doesn't exist isn't good either.

Smith said in his presentation that ROVER 6 would have National Security Agency (NSA) Type 1 encryption for Tactical Common Data Link, encrypted satellite signals, and AES 256 encryption. It will be a few years, however, until these ROVER 6 units can be deployed. As noted, it is one thing to push down a software upgrade; a hardware upgrade is both expensive and a logistic nightmare to deploy.[52]

Additionally there's the problem of simply encrypting the satellite-band video—current gadgets on the ground will be blind while those troops with the new ROVER 6 gadgets will be able to continue to rely on the signals. Providing parallel streams—encrypted and unencrypted—will only help the enemy. Fortunately, someone is thinking ahead, and the answer may be to create a secure cryptosystem, cyberlocks that may be found only in quantum physics.

6.

Quantum mechanics studies how physical elements behave in the universe. One of the properties of light is that photons can be sent in one of two modes, either vertically/horizontally or plus forty-five degrees/minus forty-five degrees. Thus, different modes can create a unique encryption key. Unlike traditional electronic and optical data transmission, quantum mechanics affords greater security since an observer can't take measurements to determine which way a photon is sent without altering it, a problem explained by Heisenberg's uncertainty principle. This alteration can be detected. Thus, any attempt to detect the signal by a third party will alter the transmission and inform the original parties.[53]

At Black Hat USA 2008, I saw a prototype for a high-speed quantum key distribution system demonstrated by the staff of the National Institute of Standards and Technology (NIST).[54] One of the challenges of cryptography, particularly when using a theoretically unbreakable "one-time-pad" encryption, is making sure that the sender and receiver—and no one else—share the same key made up of a string of random digits.[55] Quantum encryption systems use lasers to generate specific modes of photons. As noted, photons can be sent in one of two modes, either vertically/horizontally or plus forty-five degrees/minus forty-five degrees. Within each mode, one orientation could represent the digital value zero and the other the digital value one.

In one mode, a photon might be moving parallel with the ground and have a value of zero, while another can be moving perpendicular to the ground and have a value of one. In a different mode, a photon can be moving forty-five degrees to the left and have a value of zero, while another can move forty-five degrees to the right and have a value of one. The sender determines which mode and value each photon will have; the receiver chooses between modes in order to detect the signal.[56]

Using standard encryption parlance, suppose Alice is the sender and Bob is the receiver. What if Bob chooses the wrong mode and value and misses Alice's message entirely? To avoid this, the NIST system uses a wireless Ethernet channel for Alice to tell Bob what mode she used but not its value, a string of digits; Bob in turn tells Alice the values of the photons he measured. The resulting sequence, if agreed upon, is the key they both share.

In reality, Bob never sees a majority of the photons sent because Alice has told Bob how to filter the mode one way or another. He is able to decrypt Alice's intended message. For example, Alice's string might be "+/-\+-+/\" but Bob only sees "+-+-+." Bob sends Alice the values "10101," and if this is correct, this becomes their encryption key.

If another party, Eve, tries to eavesdrop, assuming she chooses the same filter that Bob did, she will destroy the original photon immediately upon interception. To keep Bob and Alice from knowing she is listening, Eve would have to generate another string to send to Bob. In doing so, Eve would introduce errors because she, too, is not seeing the entire string from Alice. If there is a significantly high number of errors, Alice and Bob can end the call and choose another encryption key, thus mitigating the MitM attack.

The system shown at Black Hat USA 2008 took a live video signal, encrypted it, and then decrypted it on the other side of the room. A simple task perhaps, but the speed and relative security were incredible: The signal was sent at the speed of light with very little signal loss. According to Alan Mink, electronic engineer in the Advanced Networking Division at NIST, the system I saw demonstrated further had to be compatible with existing fiber-optic telecom networks.[57] In particular, this technology on display could transmit a highly secure conference call between the president and his cabinet, members of which might be located anywhere in the country.[58]

7.

By now you are probably too paranoid to use your mobile phone or laptop. Don't be. You can avoid a MitM attack on your laptop in a public or private Wi-Fi space by having a firewall that detects changes in the network you are accessing. On home networks, it's important to change the default settings on your network router. This includes giving your home network a unique but not descriptive name (don't say "Susan's House" but "Squidward" instead). Then use encryption keys on your wireless network such as Wi-Fi Protected Access and Wi-Fi Protected Access 2, which means each gadget will need the encryption pass phrase in order to connect. When using public Wi-Fi spaces, such as airports and cafes, always use Secure Sockets Layer (SSL) to conduct any private transactions or use a VPN to tunnel into your corporate network. Even so, it is best not to check your bank balances or offer up your credit card too easily on such networks; attacks on wireless SSL and VPN connections are now also possible.

You can further secure your laptop to allow only connections to known AP gadgets, such as your home or office network. Set the laptop to accept connections only after a user request. For mobile gadgets, you can disable the Wi-Fi portion and risk losing the Internet where your wireless coverage is lacking. Finally, the most drastic measure would be to create a company policy that bans smart phones enabled with Wi-Fi from the premises.

If we're going to have all that cool stuff we see on TV and in science fiction, then we need to evolve networks, both wired and wireless. Specifically, networks need to accurately authenticate users and encrypt data. The quantum communications from NIST start to address these issues, but we're years away from having quantum-based personal gadgets, if not quantum-based communications.

And as we'll see in the next chapter, gadgets don't necessarily need to communicate with each other in order to betray us.

CHAPTER FOUR

Electronic Bread Crumbs

It was raining late in the afternoon in April 2007 when New Jersey governor Jon Corzine left Atlantic City in a hurry. He had been attending the New Jersey Conference of Mayors and needed to get to back to his residence in Princeton for an important meeting. That morning, radio personality Don Imus had been fired for belittling Rutgers University's Scarlet Knights women's basketball team on the air; Corzine had arranged for the team to meet face to face with Imus in the governor's office. A state trooper drove Corzine's black Chevy Suburban SUV on the Garden State Parkway, lights flashing as the vehicle wove through early rush hour traffic.

Near Galloway Township, two pickup trucks, a white Dodge and a red Ford, both drifted right to allow the motorcade through on the left. At some point, the red Ford pickup swerved left to avoid hitting a mile marker, then drove on. The action, however, forced the white Dodge pickup to collide with the governor's SUV, sending it careening into a guardrail. Governor Corzine suffered several broken bones and a facial cut.[1] Neither the trooper nor the governor's aide in the backseat was seriously injured.[2]

In the days following, witnesses, including the state patrol officer assigned to drive the governor, gave varying accounts of the accident. Most estimated that the governor's SUV had been traveling at a speed of more than seventy miles per hour in a sixty-five-mile-per-hour zone in rainy weather. Then came a state patrol report that Corzine's vehicle had been traveling at ninety-one miles per hour in the final seconds before the crash.[3] Moreover, the report explained the governor's injuries; seated in the front passenger seat, he had been the only passenger not wearing a seat belt—a violation of state law. How did the state patrol learn this? The Chevy Suburban used in Corzine's motorcade contained a black box that recorded vehicle data in the final seconds before the crash—and chances are your car has a black box as well.

The idea that your car may be—to put it bluntly—spying on your driving habits without your knowledge is very disturbing. But, as we will see, that invisible recording ability is not just within our cars. It's in our digital cameras, our photocopiers, and even those convenient tollbooth bypass gadgets on the freeway. Data is being collected surreptitiously. And the electronic bread crumbs we leave behind can lead others to reconstruct our activities years later. But can we trust the data collected in the first place?

1.

In an accident investigation, the re-creation of the steps leading to the crash is essential. Unfortunately, eyewitness accounts vary tremendously. The human mind has evolved to record only the most significant details; thus, even the best eyewitness accounts are subject to serious omissions. No one person can physically observe all the contributing factors to any one event. Not only is electronic data essential, but it may also tell us something about ourselves that we wouldn't ordinarily know. We may think we are good drivers, for example, but the data may say otherwise.

First introduced into cars by General Motors in the 1970s, black boxes were designed to measure the performance of the air bag system at the moment of a crash—and that alone.[4] Problems and deaths had been associated with early air bags' inflating too fast and too hard at the time of a crash. Triggered by the g-forces that often accompany sudden acceleration, torque, or braking action, those early black boxes, known as air bag sensing and diagnostic modules (SDMs), only recorded events in one-second intervals for five seconds before the collision. The technical data from those car crashes allowed GM engineers to reconstruct the events leading to air bag deployment and design the safer air bag system we have today.

Starting in the 1990s, GM and other auto manufacturers added memory to the SDM gadgets, allowing them to start collecting and recording data in the seconds before, during, and after a collision. Made by Bosch (formerly Vetronix), these new black boxes—they are actually silver—are known generally within the industry as event data recorders. Specifically, motor vehicle event data recorders (MVEDRs) are wired into a car's electrical system and monitor multiple sensors located throughout the car. Although the MVEDRs receive data all the time, they only record when there's an air bag deployment or an increase in g-forces.

So widespread was MVEDR use by 2005, the National Highway Traffic Safety Administration (NHTSA) decided not to require that all U.S. automobile manufacturers install black boxes because "the motor vehicle industry was already voluntarily moving in the direction recommended by the petitioners."[5] This decision later came back to haunt the NHTSA.

In early 2010, owners of various models of Toyotas reported experiencing sudden acceleration, and at least ninety-three deaths were associated with this "sticky pedal" problem.[6] In February Congress initiated a series of hearings and investigations. Toyota testified that the company did not believe the trouble to be mechanical; yet, in the media, the company suffered humiliation for not being able to explain the problem quickly and adequately.

Many of the accidents had occurred in parking lots, at speeds low enough not to trigger the MVEDRs. And part of the problem was that the black boxes were installed in only some of the cars traveling at highway speed and demonstrating sudden acceleration issues. Even then, for the cars traveling fast enough to trigger an installed black box, the federal officials had to call in Toyota officials to interpret the information recorded—Toyota does not use an industry standard format for its data.

Of the fifty-eight Toyota MVEDRs studied, federal officials found in thirty-eight cases no evidence that the brakes had been applied in the seconds before the crash.[7] If true, this supports Toyota's claim that there was no mechanical failure and implies that the driver in each of these cases mistakenly stepped on the accelerator rather than the brake.[8]

Presently, only manufacturers and some crash investigators equipped with expensive retrieval systems have access to a car's MVEDR. Car owners do not. And data is collected only in the final few seconds before a crash—not continuously, although that may soon be possible. As the price of data storage comes down, it's easy to imagine a car manufacturer recording car events over the last three or five days. This can be beneficial if a problem occurs only under certain conditions. If you take such a car into the garage, complaining about a ticking sound, with three days' worth of stored data, the mechanic will be better able to isolate the problem. Stored data could also reveal potential mechanical hazards during a routine tune-up.

Adding storage and additional data points may also help shed light on those few seconds before a crash. Automobile accidents far outnumber airplane crashes; yet, we have much more data on the latter. Such data could provide crash investigators and auto designers with better information to model the accident scene and perhaps to improve the design of future cars. In March 2010, David Strickland, NHTSA's administrator, in testimony to Congress, said his agency would evaluate the benefits of mandatory MVEDRs in all U.S. vehicles.[9]

Lost in all the discussion about whether cars should have MVEDRs installed and whether these devices should speak a common language for investigators to analyze is the question of whether this creates a privacy issue for drivers. In discussing privacy issues around these gadgets in 2005, the NHTSA's Notice of Public Rule Making Committee noted that most privacy concerns involved federal and state laws separate from the transportation agency's primary statutory authority.[10] In other words, the NHTSA punted.

What's tricky is that while the Privacy Act of 1974 allows the government to collect data about crash statistics, writes Thomas Kowalick, author of *Fatal Exit: The Automotive Black Box Debate* and cochair of the Institute of Electrical and Electronics Engineers' MVEDR working group, the government cannot engage in unwarranted invasions of privacy by using the data.[11] In crash investigations, such as the Corzine and Toyota investigations, federal investigators were bound by the Privacy Act and other laws that limit disclosure of personal information by federal agencies. Indeed, in order to gain access to MVEDRs after an accident, investigators must first obtain permission from car owners. This has its pros and cons.

Black box data could become a matter of legal record, and while the integrity of that record would need to be challenged further in the courts, this might empower others to penalize our seemingly innocent habits. An auto mechanic might be obligated by law to report repeated recordings of excessive speeds and braking or neglect of seatbelt use. With the addition of GPS data, one could also tell whether a driver was going forty-five miles per hour in a thirty-five-mile-per-hour zone on the last mile to his house. Would that be grounds for a virtual speeding ticket? Here, I think the concern is far-fetched. Today's MVEDRs only engage when there's potential for an accident; they do not monitor continuously. The return on investment for any local government to audit random cars would simply be too small to be worth it.

On the other hand, if you already know about MVEDRs, then the data they collect might be helpful to you in court. You could, for example, argue successfully that mechanical failure was the cause of an accident. Or you could compel another driver to produce MVEDR data showing his or her reckless behavior. In 2004, the *Christian Science Monitor* cited two such cases. Danny Hopkins of New York was convicted of second-degree manslaughter and sentenced to five to ten years after data retrieved from the black box in his Cadillac CTS recorded that he was driving 106 miles per hour in a crash that took the life of a young woman. In St. Louis, Missouri, Clifton McIntire pleaded guilty to manslaughter after data from the black box in his GMC pickup showed him driving eighty-five miles per hour before slamming into a Toyota and severely injuring a young woman.[12]

Kowalick said privacy issues affect not only who owns the data but also how it's collected, processed, transmitted, and stored.[13] In the past the auto industry has used MVEDR data to improve the safety of its products, but the automotive insurance industry has expressed interest in accessing MVEDR data as well. Instead of relying on a driver's word or on established claims, an insurance company would remotely access your car's data, which, in theory, would allow it to adjust your specific insurance rates accordingly. One proposal in the California legislature would have allowed insurance companies to charge members only based on the number of miles actually driven within a billing period. Such a pay-as-you-go system could use GPS data supplied by the car's black boxes.[14] This financial incentive may also improve our general behavior on the road. However, few people will be comfortable knowing a third-party commercial interest is monitoring their specific driving habits.

This raises long-term questions regarding such data. If insurance companies want access, who else might benefit? One possibility would be data warehouses. Imagine a scenario where a data warehouse includes, among other data, your driving record. This could impact your ability to get a good rate on your next auto loan. It might also influence health insurance (under the category of risk factors, driving recklessly could be counted against you). Perhaps this is one area where federal regulation should play a direct role.

But can MVEDR records be trusted? In some court cases, the data collected from the MVEDR has contrasted with conventional crash forensics and credible eyewitness accounts.[15] In 2004, the governor of Maine, John Baldacci, was involved in a serious accident on I-295, near Bowdoinham. As in the Governor Corzine incident in New Jersey, a state trooper in Maine, who was responsible for driving Baldacci, claimed he was driving a reasonable speed, in this case fifty-five miles per hour, when he attempted to pass a slow-moving vehicle.[16] The MVEDR in the car, however, reported the vehicle traveling at seventy-one miles per hour at the time of the crash. Traditional forensics, however, supported the state trooper's account, explaining that the wheels spinning on the ice could explain the higher speed recorded by the MVEDR.[17]

A second dispute arose about whether Baldacci was wearing a seat belt at the time of the accident. The MVEDR recorded that he was not. This the governor vigorously denied, and the emergency room doctors even testified that his injuries were consistent with seat belt use. Ultimately, Maine's public safety commissioner ruled the MVEDR was inaccurate since "the clear and convincing physical evidence and the interviews of the involved parties were sufficient to satisfy the questions raised by the conflicting data and it is the State Police conclusion that Governor Baldacci had his seat belt buckled."[18]

An MVEDR can only report on internal data—what its internal sensors tell it—not external circumstances. A similar ruling occurred in a case in Maryland. There, a woman was charged with driving recklessly and crossing the center divide. The charges were based solely on the data from her car's MVEDR. Again, the black box failed to take into account patches of ice on the roadway at the time.[19]

Unless you're in a serious accident, chances are your driving data will remain private; if you have been in a serious accident, that data has probably been used to help make new cars even safer. Deep down though, this assurance does little to quell our unease about potential violations of privacy and raises several interesting questions: Who really owns the data? For how long should the records be kept? And, most importantly, how accurate is that data?

As we will see, common gadgets today collect all kinds of information—some we can access, some we cannot—and this information, regardless of whether the gadget is accurate, is being used, sometimes years after the original collection, in courts of law.

2.

The traditional and most common use of black boxes—indeed, the one that we're all familiar with—is in the airplane industry. In the wake of the mysterious crash of a French airplane over the Atlantic Ocean in

2009, there were calls to have airplanes provide real-time broadcasts from these black boxes.[20] Under this proposal, planes flying over the ocean or polar caps would use the global Iridium satellite network to provide in-flight data to ground stations. The satellite-based gadgets are, however, five to seven times more expensive than conventional black boxes. Additionally, the volume of data broadcast by the two black boxes on commercial jet planes requires satellite data costs of $3 to $5 per minute. To minimize these costs, vendors of the AMS and Star Navigation systems say they can switch to live broadcasting only if there's an incident. The pilots can also broadcast live data if needed.[21]

Something similar is happening with black boxes in cars. On the ground, the use of satellite signals is not necessary; simple cellular communications are enough to collect data remotely in real time.

A Mercedes commercial in the United States depicts a calm voice from the dashboard of the family car informing the driver, a middle-aged mom, and passenger that the oil gauge is getting low, and, by the way, it is also time for a tune-up. Using hands-free communications, the woman, with her cute child strapped into the backseat of the car, calmly makes the maintenance appointment without ever taking her eyes off the road. The commercial is meant to stress the convenience and safety of having the auto manufacturer keep track of mundane maintenance issues.[22]

At first the commercial makes it sound as though the manufacturer is drawing upon static records ("it's been six months since your last oil change"); actually, telemetric sensors (combining computer data and communications) within the Mercedes beam live data to a central database, where that data is flagged by a customer-service representative, who then calls the customer to make the appointment. Often the local dealership receives the car's diagnostic information ahead of an appointment.[23]

Motor vehicle data is no longer passive or localized solely within the black box inside your car. With telemetrics, it is active, sometimes transmitted directly to the auto manufacturer in real time and monitored remotely. As long as the data is diagnostic, most people should see this as simply enhancing vehicle safety.[24]

After test-driving a 2005 Chevy Malibu Maxx, the editors at *Auto-Week* wrote of a strange conversation the driver had with a representative from OnStar, a subscription-based monitoring service included with the purchase of the car.[25] The editors, who routinely drive a car hard in order to make competent statements in their product reviews, had just taken the SUV around a particularly challenging obstacle course. Shortly after they completed the maneuver, their OnStar button lit up. The voice on the audio system asked if everything was all right. No one in the car had reported a problem; nor had the car crashed. Rather, the vehicle's many sensors had kicked in because of the increase in g-force data and reported that event immediately to OnStar.

OnStar, the creation of General Motors, Electronic Data Systems, and Hughes Electronics,[26] offers drivers many safety and security features, such as Automotive Location Identifier, which sends GPS coordinates directly to emergency services in the event of an accident.[27] The technology also allows OnStar to rate the severity of each crash based on various data points. This is designed to assist subscribers too injured to speak for themselves. OnStar has partnered with Verizon cellular to offer hands-free mobile phone service. Using a cellular network, OnStar has more features, such as the ability to lock and unlock an OnStar-enabled car remotely, contributing to a customer's peace of mind.

More controversial is OnStar's ability to decelerate a speeding car remotely.[28] In theory, the victim or the police would first identify a stolen or speeding vehicle, then have OnStar send a kill signal to the car. OnStar has said the kill signal, known as Remote Engine Block, will only be sent when police are within eyesight of the car. The kill signal tells the accelerator to ignore any signal from the driver, so the stolen or speeding vehicle will eventually slow to an idle. The ability to hack this signal is suggested in the 2008 film *Untraceable*; however, gaining the resources to hack Verizon's CDMA cellular signal (which we defined in Chapter 3 as more complex than GSM) would require the backing of a wealthy nation and thus would be unlikely to affect the common OnStar subscriber.

Like that captured by a black box, OnStar data provides a valuable record of the driver's behavior. OnStar's data privacy policy states that the company keeps all records for one year, but the company doesn't say what happens beyond that period.[29] Even if OnStar does remove all the data after one year, another record is still being kept: the mobile phone data. For billing purposes, mobile phone carriers keep detailed records, some going back several years, but they lack a privacy statement similar to OnStar's. As we'll soon see, this record of which cell towers you used when and for how long could one day come back to haunt you.

3.

Technology reporter, commentator, and former TechTV television personality Cat Schwartz learned firsthand how gadgets—in this case, digital cameras that store the date and time and even the original image—can betray us. The former star of *The Screen Savers* and *Call for Help*, Schwartz once posted some professional photographs of herself cropped in odd shapes on her website. Fans wondered what had been removed from the photos. Given that these were technology fans, some knew to look within the Adobe Photoshop image file itself. There, along with the minutia of shutter values and digital colors, were the uncropped nude photos of the TechTV star.[30]

Consider how we use our digital cameras today. Most often we shoot a picture on our phones or with a digital camera and upload these images to the Internet. Online digital photo albums are great for sharing images with loved ones, but digital cameras, unlike their film counterparts, automatically imprint a lot of extra data in what's called exchangeable image file format (EXIF) embedded within each image file. Some camera manufacturers also use a similar format called raw-image format, which also includes a variety of data from the camera. Either way, data is collected that you often don't see or think about.

Unless you know how to remove this information later with image-editing software on your PC, the EXIF information will likely remain with the image file long after you post it online. What harm is there in that?

If you never crop or otherwise alter the photo, EXIF poses minimal risk and numerous benefits. EXIF allows online albums to chronologically order your images and printers to reproduce the color and light conditions accurately. But if you crop an ex-girlfriend out of the picture, the preview image with girlfriend will sometimes remain within the digital file itself. This is like exposing redacted information in a Word document stored on a government website.

Here the added convenience of having archival information within our gadgets means we need to adapt our behaviors constantly to our use of these technologies. If you think it's safe to post any old image online, you might want to think again.

Photos uploaded online today, especially those taken with our mobile gadgets, have the ability to identify the date and time, as well as the exact spot where the photographer stood as GPS data is incorporated into the image file. Years from now, that information might be valuable, say, in re-creating a classic photo. But most of us are not-so-great photographers, and we need to consider the unintended ways this information might be used. What if you tell your boss you're taking a couple of sick days, when photos online reveal that you were actually attending Lollapalooza. And that is just the beginning.

Online and digital photos are proliferating. Sites like Twitter and Facebook allow account holders to upload and share pictures. And gadgets like the Apple iPhone, BlackBerry, and Google Android automatically take images with geolocation-enabled EXIF. Ben Jackson of Mayhemic Labs noticed this too.

Although their primary goal is to "do cool stuff" in vulnerability and malware research, the staff at Mayhemic Labs may be best known for creating a site called ICanStalkU.com. Random tweets that contain images are reposted along with links to a map location, the original tweet, the tweet image, and the ability to send a reply to the person.[31]

For his presentation at Next HOPE (Hackers on Planet Earth), a security conference in New York City, Jackson used the information learned from adult-themed images from SexyPeek.com to illustrate what he could learn about anyone posting almost any image online. Through public records, Jackson was able to find a name associated with the house at the latitude and longitude where the photo was taken. The owner, however, produced no online presence. Using Google, Jackson discovered more geotagged images that could all be traced back to the same BlackBerry 9000 that had posted the original image, some at the same longitude and latitude as the house but posted online under a different name. Using that different name, Jackson found a Facebook account with a birth date, marital status, and friends. The man even had a second Facebook account under yet another name. This person was clearly keeping secrets, a fact his wife had discovered herself, writing on her Twitter account that her husband "has more secrets than I've ever guessed, guess that's why he thinks I'm always hiding something, cuz he's hiding stuff."[32]

All of this Jackson learned from just one photo. He has collected thousands of such pictures. At Next HOPE Jackson made public his database of information gleaned from TwitPic.[33] He admits not knowing what can be done with the information, other than showing people just how much information is being leaked by one single image posted to a public site. Specifically, Jackson's file contains the latitude and longitude for over 5,000 TwitPic, yfrog, and SexyPeek URLs.

At Next HOPE, Jackson also released two tools. One, called Reaper, allows others to take a URL from a service, extract the EXIF tags, and look for geotagging. Another, called Stalker, scans a single user's entire photostream on TwitPic or yfrog and outputs the data to the console or to a file that can be imported to Google Earth.

Hopefully this will make you think twice about uploading or sharing that digital image. Unfortunately, universally disabling geolocation on iPhone or Android phones will break geolocation across all your applications; however, Jackson notes, it's possible to disable the "store location" feature only on the phone's camera.[34] But who among us ever takes the time to do this with each gadget we own?

4.

Getting rid of old computer equipment is no different from tossing out an old sofa—or is it? Putting an old PC out by the curb for the sanitation workers to come and take away might seem practical, but it's not a good idea. At least take out the hard drive before recycling old computer equipment.

Whatever you do, don't then sell the hard drive without first making sure all the data has been removed. A study by forensic firm Kessler International found that four in ten used hard drives purchased on eBay contained usable data.[35] They reported finding within their sample group that

36 percent had personal and confidential documents, such as financial information;

21 percent had e-mails;

13 percent had pictures;

11 percent had corporate documents;

11 percent had Internet browsing histories;

4 percent had Domain Name System server information;

4 percent had miscellaneous data.

The ability to recover data from secondhand hard drives is not new; I first wrote about the dangers back in 2001.[36] The Microsoft Windows operating system is notorious for not really deleting data, which makes it recoverable later. When you hit delete, the file is removed from the master file record, but the deletion is often trivial, merely removing the first character of the file name. The contents of the file remain; they just can't be found easily. Thus, a file called "deleted" looks like this: "_eleted." Recovery software can replace the first letter, then reconstruct the master file record to include these deleted files. That's how deleted Enron e-mails, as well as lost files from the George W. Bush administration, were recovered.[37]

A data file is composed of several clusters. When a file is saved, the data first fills up one cluster and then, if necessary, spills into another. The clusters need not be adjacent to each other. Deleting a file removes all those cluster locations from the master file record, and the drive frees them up to be written over with new files. Often they are skipped and other, virgin clusters are used.

In cases where clusters are rewritten, the new data may not fill the entire cluster; the old file with the old data still peeks through at the very end of the cluster.[38] There are ways to reconstruct this "slack data," the bits and pieces of old files that haven't yet been overwritten. And simply writing new files over the old isn't good enough; techniques exist to recover files that have been overwritten as well.

What works is to rewrite the old drive with layers of ones and zeros, or "noise." The U.S. government requires that discarded hard drives and related media be rewritten seven times—that means adding seven layers of ones and zeros so that someone doesn't come along later and try to reconstruct the drive. Many commercial products are able to "scrub" or "shred" the drive up to a hundred times, ensuring that the drive is thoroughly wiped clean. Most people never take the time to do this; yet, the benefits of digital (and paper) file shredding are clear.

And it's not just hard drives on old computers that pose a long tail of risk. People also sell old digital cameras, MP3 players, and media cards that still contain data. For media cards used in digital cameras, several free programs on the Internet are capable of data recovery. The best, such as Forensic Toolkit and IDA Pro, cost thousands of dollars, but if you have a used media card, several free programs available online will at least identify whether there's any usable data on the card.

In an InformIT study, a research team bought sixteen used media cards on eBay. Of these, only fourteen were ultimately available for analysis. One contained 226 photos that had been deleted but were not really gone; they could be recovered. Some of the media cards contained images viewable without any forensic recovery whatsoever.[39]

For most people, it's not a problem whether someone discovers their passion for all the remixes of "Achy Breaky Heart" or their photos of curious cloud formations. But given the existence of EXIF data, there's the added possibility that someone might use the GPS data within the recovered images to track you down.

Even if you don't sell old hard drives or media cards on eBay, you may still have personal data in places you might not have realized. Just in time for tax season in 2007, Sharp sent out a press release using a Roper survey of 1,005 adults concerning the security of digitally processed information; it warned that "a majority of Americans are not aware that their personal and financial information may be at risk for identity theft."[40] The point of the press release and survey was that commercial copiers, like those sold by Sharp and other vendors, today contain hard drives, and without proper security methods, the documents you photocopy at a commercial photocopy shop may be retained on the machine long after you leave the store.

What if you have written a rather large manuscript and want to copy it at a local copy shop. The photocopy machine will scan onto a hard drive the majority of the pages while printing the first few. This saves time, but it also leaves behind on the hard drive a copy of the manuscript, which is your intellectual property. More applicable to you and me, the Roper survey found that more than half of the respondents made photocopies of their tax returns, and almost half of those people used commercial photocopiers at a shop or public photocopiers such as those at the local library. When servicing the machine, the technician, if so inclined, could access the documents on the drive. More likely, the technician might need to replace the drive, and the old drive might simply end up on eBay. "You actually have a better chance at winning 10 straight rolls of roulette than getting those hard drives on copiers rewritten," said Ed McLaughlin, president of Sharp Document Solutions Company of America, which now encrypts data on its photocopier hard drives.[41]

This caveat also applies to digital photo kiosks. After you insert a media card or USB drive, the kiosk may print out your picture, but it might also store the image as well as, if your camera captures it, the EXIF data. An alternative would be to upload the images to a website for printing; at least then, you know who has a copy of your image.

5.

Today we're unconsciously leaving electronic bread crumbs wherever we go—in our cars, in our phones, in the hard drives of gadgets we may not have considered before. But what about systems we consciously opt into? How aware are we of the long tail of risk when the data is generated from an activity we want? When data that we let others collect is spread out over several years, seemingly innocent and random trips can later, when viewed in aggregate, produce unintended consequences.

Electronic tollbooth transponders are changing how we commute. Instead of waiting in long lines to toss loose change into a metal bucket, some drivers simply whizz through an open lane where electronic readers record the car's unique ID number in order to bill an account at the end of the month. Subscribers to these tollbooth systems, marketed under the names E-ZPass, FasTrak, and I-Pass, among others, voluntarily give transit officials remote access to the miniature circuit boards mounted inside the windshield of a registered car. The gadgets use radio frequency identification (RFID) to broadcast their data over short distances to readers located within tollbooths and on the road itself. Monitoring these transponders allows transit officials to measure the time taken to cross a bridge, to go from tollbooth to tollbooth, or to move from roadside monitor to roadside monitor.

Nate Lawson, a Bay Area independent security researcher, has discovered, despite public claims to the contrary, that the unique ID used by FasTrak and other transponder systems isn't encrypted; the IDs are transmitted unencrypted for anyone with the right equipment to see. Even though the toll transponder doesn't contain personal information, its unique ID still holds value. If someone stole your transponder's unique ID, would that person then be able to use a toll road on your dime? They could, said Lawson.

To make matters worse, Lawson said he could both read and write the new data to his tollbooth transponder—something else that should not be possible. Thus, he could enter any Bay Area parking lot, use a specially configured laptop to emulate a tollbooth reader, query the available transponders, record their unique IDs, and later clone someone else's ID onto his own transponder. He could then use the cloned transponder to enter San Francisco for free (Bay Area tolls only charge to enter San Francisco; you are always free to leave). Informed of Lawson's finding in mid-2008, the Bay Area Metropolitan Transport Commission said the privacy and security policy of the overall system was being reevaluated.[42]

You could possibly defend yourself against high toll charges by requesting the license plate photos taken at tollbooths. All autos, whether part of the transponder program or not, have their license plates photographed at tollbooths.[43] In the case of someone's failing to pay, a human being at some later point will look up the scofflaw's license plate photo. Once a human matches the license, the driver receives a bill.

One consequence of cloning is that the transit district still makes its money. The user whose transponder ID was stolen might be charged extra at the end of the month—unless the user is vigilant (which most of us are not) and looks over each and every charge (as one should do with every bill or credit statement). However, if the user does contest the charge, then the district loses the money. Over time, cloning could become a serious problem for any transit agency (and in the next chapter we'll look at cloning transit cards in greater detail).

Records of tollbooth transponder account activity are kept "indefinitely." The San Francisco Municipal Transportation Agency, for example, said that in addition to toll transit times, its system keeps a record of a registrant's credit card number and license plate photos. They do so to allow drivers to dispute charges or violation notices years after the fact. "We've had cases where people come back a year or more after and say, 'This is a wrong charge,'" a spokesperson told the *Oakland Tribune.*[44]

Unlike mobile phone records, transponder records are relatively easy for a lawyer to obtain. And there is virtually no case law around the privacy of such records. Neither the privacy policy nor the customer license agreement for FasTrak or E-ZPass explains that court orders may include subpoenas to obtain transponder data in civil cases, which do not require a judge's approval unless they are contested.

For example, let's imagine an office building saves on its energy bill by turning on specific hall lights only when you walk by—the light follows you, but the hallway remains dark before and after your passage. This could be accomplished using nontracking motion detection. Or the lights could be triggered by the RFID tag on your employee work badge (we'll talk more about RFID and work badges in the next chapter). The difference is that motion detection is anonymous; the work

badge version can create an audit log or record. Using the unique identifier in your badge, someone could go over the logs and determine your
movements through the office during a given day. Suddenly the innocent act of going to the washroom at midday could be timed and later
deemed excessive by an overanxious manager.

You're probably thinking that if you have nothing to hide, then why
should it matter whether transit agencies track and retain a record of
your habits for future use? The fact is that others may not always see
you as such a fine individual.[45]

In 2007 a *Contra Costa Times* newspaper investigation revealed that
twenty subpoenas had been served within the previous twenty-four
months on some of the half-million FasTrak users in the San Francisco
Bay Area—an average of about one subpoena a month.[46] Law enforcement investigations account for a handful, but not all, of the requests. The rest were for civil suits, such as an employer using the toll
transponder records to prove that a work-at-home employee wasn't at
home but driving across one of the Bay Area's toll bridges during work
hours. This data has also been used in divorce proceedings: "Part of the
reason Fred has not had success . . . is that he takes too much time off,"
claimed one woman who sought her husband's toll activity in one divorce case. "His transponder records . . . will show how little he works."[47]

This behavioral profiling is not only a West Coast occurrence. According to a 2007 Associated Press survey, agencies in seven of the
twelve states in the Northeast and Midwest that are part of the E-ZPass
system have provided electronic toll information in response to court
orders in criminal and civil cases, including divorces, with some re-
quested logs going back several years in order to prove a pattern.[48]

While we might say we didn't cross a tollbooth at a given time, an
electronic record still proves that we did. But what if we really didn't
cross the toll plaza at that time and can produce witnesses who place
us in another location? Like Governor Corzine's and Governor Baldacci's black box data, tollbooth transponder records might not yet be
accurate enough to stand on their own.

In a few court cases, tollbooth transponders have factored into the
final verdict. For example, toll transponder logs were used in Melanie

McGuire's 2007 murder trial in New Jersey. Defense attorneys maintained that William McGuire was heavily in debt from gambling and likely killed by someone else. The prosecutors, however, believed differently and used mobile phone, transponder, and GPS records to build a case against his wife. By researching E-ZPass records, investigators learned that after his death, someone took William McGuire's Pathfinder on a 1 a.m. drive to the Chesapeake Bay. Not once, but twice. Perhaps not coincidentally, McGuire's hacked up body was found in suitcases floating in the bay not long afterward. Investigators then asked Melanie McGuire about the use of her husband's vehicle on the nights in question. Shortly after the interview, an unidentified female called the E-ZPass office to dispute the billing record. In court prosecutors used mobile phone records to show how the billing-dispute phone calls actually came from Melanie McGuire. In the end, the mobile and transponder records were integral to the conviction.[49]

If we're going to keep digital records and use them in court someday, how accurate are they? Perhaps not as accurate as you might think. With a proliferation of data, it's possible to commit virtual identity fraud by cloning electronic ID data. And as we'll soon see, sometimes it's possible to obtain someone's full identity (name, address, and Social Security number) without really trying.

6.

At Houston's, a California restaurant chain, twenty-nine-year-old Jem Matzan ordered a veggie burger with some drinks and was asked to provide his driver's license. Matzan handed over his Florida driver's license, only to find the waitress disappear and reappear moments later with his card and the drinks. When he asked why she'd taken the card, the waitress explained, "We scan it through a machine to make sure it's real." Matzan threw a fit, but by then it was too late.[50]

What harm is there in having your driver's license swiped? On the face of it, most of us instantly know not to give away our driver's license; it's the closest thing we have to a national identity card. Besides a passport, it's the only piece of information that identifies us at the airport, for example. Allowing the bar code and magnetic strip on the backside of a driver's license to be scanned is like physically giving our license to someone else to own. Unlike photocopies of our licenses, which are routinely submitted for preschool pickup privileges, a swipe of the card makes the data digital and thus easy to store in a database. Matzan provided far more than name and address or date of birth; he also gave his height, weight, eye and hair color, and digital forms of his image, fingerprint, and legal signature. We might be okay with having that data added to a government database, swiped in by a law enforcement officer; they have most of our information on record anyway. But if it were entered into a commercial database—owned by the liquor store down the street—would we be okay with that?

Paul Barclay, owner of The Rack, a former Boston nightclub, told the *New York Times* in a 2002 interview that he had collected the personal information of over 1.3 million patrons using a commercially available Intellicheck scanner.[51] In addition to purchases by retail establishments, Intellicheck Mobilisa sells its ID Defense scanners to the U.S. government and military. The company's chief executive officer, Dr. Nelson Ludlow, has suggested, instead of requiring people to carry a separate work ID for access to a building, why not simplify life and use the card that everyone has with him or her all the time, a driver's license?[52]

Some models of Intellicheck are capable of running background checks. The gadgets are expensive—starting at $2,000—but Barclay claimed the scanner helped block repeat offenders from entering his nightclub. Persons known to be troublemakers or underage could be bounced after their cards were swiped at the bar. In 2005, Barclay told the *Times* that the scanner helped him confiscate six hundred fake IDs, thirteen in one weekend alone at his nightclub's former Faneuil Hall location.[53]

At stores like Target or Payless ShoeSource, the Intellicheck ID Defense swiper is used both to ring up sales and to swipe driver's licenses for enrollment in the store's loyalty or store-branded credit card programs. Intellicheck Mobilisa states on its website that more than eighty banks currently use the swiper to open new accounts. And along the

U.S.-Canadian border, the company claims the gadget saves U.S. Customs officials time by swiping driver's licenses (although, as we'll see at the end of Chapter 5, that claim is dubious).

Even Barclay used the scanner for purposes beyond just bouncing underage people at the door. For example, the information Barclay collected at The Rack told him that on Tuesdays, when jazz performers regularly performed at the bar, he could count on a spike in customers born between 1955 and 1960. The data also revealed that another night of the week was popular with customers from selected zip codes in the Boston area. And the use of driver's license information works both ways. Law enforcement has already called on bars such as The Rack to see whether certain names and Social Security numbers have shown up in their databases. "You swipe the license," Barclay told the *Times*, "and all of sudden someone's whole life as we know it pops up in front of you. It's almost voyeuristic."[54]

Since they were first created in 1933, state-issued driver's licenses have always varied in information and detail from state to state, with some states printing more information and others less. In the past, this made traveling to another state and getting pulled over by a cop an interesting experience as the officer attempted to locate basic information on an unfamiliar card. That changed in the early 1990s, when the American Association of Motor Vehicle Administrators (AAMVA) introduced what it hoped would become a national standard. The association sought to regulate what should be included on all state-issued driver's licenses. The AAMVA recommendations, for instance, called for the same ISO 14443 magnetic strip standard used on credit cards, then a one-dimensional bar code, and later a two-dimensional bar code known as PDF417. The machine-readable magnetic strip and bar code would not be encrypted. These two recommendations—machine readable and not encrypted—have had a profound effect on privacy. Not only can law enforcement officers in different states swipe a driver's license and glean all the relevant data, but so can commercial establishments. Only Nebraska, New Hampshire, and Texas prohibit the swiping of driver's licenses for commercial uses, but as we will see, even within these states there are exceptions.

It's not the first time driver's license information has been ridiculously easy to obtain. Up until 1994, a state's selling Department of Motor Vehicles (DMV) information directly to marketers and others was common and perfectly legal.[55] For example, before 1994, clothing manufacturers would buy DMV lists in order to target catalogues to people of specific heights and weights in specific zip codes. Unfortunately, it took a tragedy to change that.

In the summer of 1989, nineteen-year-old Robert John Bardo wrote letters to celebrities such as Samantha Smith, Deborah Gibson, Dyan Cannon, Madonna, and Tiffany.[56] Bardo, who grew up in Tucson, Arizona, is the youngest of seven children and at the time was a janitor at a local hamburger stand.[57] That summer he started writing long and obsessive letters to Rebecca Schaeffer, twenty-one, then costarring with Pam Dawber in the hit television show *My Sister Sam*. Newly successful, Schaeffer was breaking into film and had recently appeared in *Scenes from the Class Struggle in Beverly Hills*. On the day of her death, she was preparing to audition for a part in *The Godfather, Part III*.

Bardo had previously traveled to New York to visit the spot where Mark David Chapman shot John Lennon. He had also read a *People* magazine article about Arthur Jackson, a stalker who acquired the address of Theresa Saldana, an American actress who played Joe Pesci's wife in *Raging Bull*.[58] In 1982 Jackson called Saldana's mother and posed as a director's assistant; he later showed up at Saldana's house and assaulted her. She survived.

Apparently inspired by Jackson, Bardo paid an Arizona private detective to obtain Schaeffer's home address from the California DMV, and on July 18, 1989, he flew to Los Angeles. According to court records, he made two visits to Schaeffer's home that day. First he rang her doorbell, had a polite conversation with her, and then left. But in leaving he realized he had intended to give Schaeffer a CD, so he rang again. According to Bardo's own testimony, Schaeffer seemed a bit annoyed that he'd returned, and her tone led him to shoot her at point-blank range with a gun one of his brothers had purchased for him. (Incidentally, the prosecutor for the state in *California v. Bardo* was

Marsha Clarke, who four years later would be the lead prosecutor in the O. J. Simpson trial.) In 1991, Bardo was sentenced to life without the possibility of parole.

As a result of Schaeffer's murder and Saldana's advocacy for anti-stalking legislation, the state of California passed a law making it illegal to obtain anyone else's personal information through the DMV.[59] A similar bill was introduced to the U.S. Senate by California's senator Barbara Boxer and cosponsored by California's other senator, Dianne Feinstein, along with a dozen others.[60] Various California representatives sponsored a House version.[61] In 1993 President Bill Clinton signed into law the reconciled legislation, known as the Driver's Privacy Protection Act, and as of 1997, information collected by a state's DMV for the purposes of issuing a driver's license is protected by federal law.[62] Still, there are exceptions.

As part of the USA PATRIOT Act's renewal in 2005, California's Feinstein added the Combat Methamphetamine Epidemic Act (CMEA), a law that requires pharmacies to restrict the sale of the decongestant drug pseudoephedrine and certain other drugs that are, collectively, used in the production of methamphetamine.[63] Apparently Feinstein didn't envision the compromise to personal privacy that would follow. The law as written requires pharmacies to keep a photo ID on record for up to two years with any purchase of pseudoephedrine, the active ingredient in the decongestant Sudafed. As a result, a modified version of Sudafed (Sudafed PE using phenylephrine) is now sold over the counter. Nowhere, however, does CMEA explicitly state that the photo ID must be a driver's license; it says only, "Seller may not sell the product unless prospective purchaser presents a photographic identification card issued by a State or the Federal Government, or United States passport (unexpired or expired), Alien Registration Receipt Card or Permanent Resident Card, among other acceptable forms."[64]

To obtain the drug pseudoephedrine, a customer needs to show a government-issued photo ID and fill out a paper logbook. Yet, many national pharmacies—out of convenience—have chosen to install commercial driver's license swipers, such as those offered by Intellicheck

Mobilisa, as a quick means of satisfying the law. One drug manufacturer sells a license scanner made by Honeywell using this very argument: "Many retailers use paper-based logbooks to comply with the CMEA. However, a paper system creates challenges for everyone involved. It is time-consuming for consumers and retailers to fill out and complete; presents risks to retailers if logs are not secure and accurate; costs retailers money to maintain logs; and is difficult and time-consuming for law enforcement to enforce the mandate."[65]

When confronted with a sick child, many of us will swipe anything that allows us to get that medication home quickly. In that case, it's probably okay: It is a matter of trust. We trust our pharmacists to protect our privacy. After all, they have to provide the correct dosages and bill our insurance. We don't, however, have the same incentive to trust our local restaurants, bars, or convenience stores.

7.

As we will see in Chapter 7, data collection with the personal information removed can be used for the common good. Such sanitized data can yield vital information about our collective habits without compromising privacy. But there is movement in the opposite direction, a decision to collect as much personally identifiable information as possible and to store that information for an indefinite period in order to fight terrorism.

In 2009, the National Security Agency (NSA) announced the construction of a data center in the Utah desert to house "signals intelligence." Such a data center would provide for the analysis of various forms of communication—phone calls, e-mails, and other data trails such as Web searches, parking receipts, bookstore visits, and other "digital pocket litter."[66] Given the mistakes we make interpreting small collections of data—black boxes in cars and toll transponder records—one can only imagine the errors made possible by collecting even more data from an even wider variety of gadgets.

Writing in the *New York Review of Books*, security expert James Bamford notes what many of us suspect is true:

Just how much information will be stored in these windowless cybertemples? A clue comes from a recent report prepared by the MITRE Corporation, a Pentagon think tank. "As the sensors associated with the various surveillance missions improve," said the report, referring to a variety of technical collection methods, "the data volumes are increasing with a projection that sensor data volume could potentially increase to the level of Yottabytes (10^{24} bytes) by 2015." Roughly equal to about a septillion (1,000,000,000,000,000,000,000,000) pages of text, numbers beyond Yottabytes haven't yet been named. Once vacuumed up and stored in these near-infinite "libraries," the data are then analyzed by powerful infoweapons, supercomputers running complex algorithmic programs, to determine who among us may be—or may one day become—a terrorist. In the NSA's world of automated surveillance on steroids, every bit has a history and every keystroke tells a story.[67]

Such a data-collection scheme has been tried before—with disastrous results. In the 1930s, the Netherlands put into place a comprehensive registry of name, birth date, address, religion, and other personal information about each resident, no matter what age. The idea, very simply, was to improve the efficiency of government and public welfare planning. Then the Nazis invaded. Suddenly, a hostile power had access to the personal information of everyone in the country. Thus, the Nazis were able to filter the population to its liking. Unlike in Belgium, where 40 percent of the Jewish population was murdered, and France, where 25 percent of the Jewish population died at the hands of the Nazis, according to Viktor Mayer-Schonberger, director of the Information + Innovation Policy Research Centre at the Lee Kuan Yew School of Public Policy, National University of Singapore, 75 percent of the Dutch Jewish population died because of this complete national registry. On the other hand, Jewish refugees from other nations who lived in the Netherlands at the time survived, in part because they were not required to register. The Gypsy population was hit hardest and almost completely eradicated from the country by the Nazis.[68]

It isn't just government agencies that are collecting random information. As we'll see in the next chapter, the rush to broadcast personal data such as our driver's licenses via RFID and related technologies means that just about anyone can not only amass a detailed record of our comings and goings but clone and misuse our personal information.

CHAPTER FIVE

Me, I'm Not

In the summer of 2009, researcher Joe Grand figured out how to get unlimited free parking within the city of San Francisco. Working with fellow researchers Jacob Appelbaum and Chris Tarnovsky, Grand had turned his attention to the 23,000 "smart parking meters" being installed around the city as part of a $35 million pilot project. One flaw they discovered allowed them to increase the value of the plastic one-time smart cards used for payment—for example, Grand created a counterfeit card worth $999.99. He also found he could freeze that maximum value—not have it decrease with each use—and thus make his one-time-use card eternal. This would give him free parking in San Francisco forever—except Joe Grand isn't a thief.

Grand, known as "Kingpin" within the security community, is a youthful old-school hardware hacker. As a teenager in Boston, he was a member of L0pht Heavy Industries, a hacker "collective" most recognized for discovering early vulnerabilities in Microsoft Windows and testifying before the U.S. Senate in 1998 that they could shut down the Internet in thirty minutes. L0pht also produced several security advisories. They created software tools, including L0phtCrack, a password cracker that is still in use today. In 2000, L0pht merged with security consulting company @stake, which Symantec acquired in 2004. Grand, however, went independent and stayed with hardware hacking, then a relatively unknown skill. He eventually moved west, settling in San Francisco, where his Grand Idea Studio today consults with industries on various aspects of security and product design. He has cohosted *Prototype This* on the Discovery Channel and remains active within the security community, exposing vulnerabilities in hardware technologies.

Some of the new credit card–based meters owned by the Port of San Francisco and used along the Embarcadero are wireless; however, not all of them link together wirelessly in a grid, perhaps because the city wants to retain its army of meter collectors. To collect the fares, these workers zap each meter with a specialized PDA. So, every few days, a meter collector comes along, takes any coins from the parking meters, and uses a specialized PDA to record audit logs and make any programming fixes.

"Every electronic-based system needs a legitimate way for a maintenance worker or meter maid to communicate with it, usually for retrieving audit logs or diagnostic information," said Grand.[1] "So there are legitimate ways in, and if there are legitimate ways into the meter, then you could use those legitimate ways in for a possible attack." Grand wanted to analyze San Francisco's PDA system. "I have seen a meter maid use one [PDA]," he said, "but I haven't been able to touch one or use one myself." Instead, Grand focused on the smart card itself, which contained a tiny microcontroller, and the way that it communicated with the parking meter.

The parking smart cards are available in amounts of $20 and $50 from local San Francisco merchants and cannot be recharged; once the value is depleted, they are discarded. Grand says that for his experiment he paid cash for ten $20 cards. When the merchant asked Grand why he needed so many cards, he explained he was a sales manager and wanted his team well supplied. (Unsure of how the city of San Francisco would respond to his research, Grand didn't want to tell the merchant he was looking for a way to break the system.) Because each card has a unique serial number, in theory the local parking authority could track your movements, or at least the movement of a card through the system. However, the ability to pay cash for the smart card and the fact that no one writes down the serial number keeps the identity of the actual user anonymous.

The goal for the team was to learn how the meters in San Francisco interact with smart cards and then look for flaws they could exploit. Grand and Appelbaum first purchased a few used older versions of MacKay parking meters on eBay. By opening up these systems and examining their electronic guts, they got a general idea of how a parking meter is designed without having to chop one off the street. They still, however, needed to see exactly how San Francisco's implementation worked. So, Grand and Appelbaum wandered San Francisco with a portable oscilloscope and special shim card that fit into the parking meter's smart card socket. Their setup—essentially a digital form of eavesdropping—allowed them to monitor and capture all electronic communication between the card and the meter. It looked odd, Appelbaum said, standing there on the curbside with a circuit board jammed into a parking meter, wires trailing out; yet, he joked, "In San Francisco, if anyone ever asks, you explain, 'It's for an art project,' and no one will think twice about it."[2]

In just three days, using nothing more than a pen and paper, Grand managed to decipher enough of the communication between the smart card and parking meter to figure out how value was deducted from the card. In the following days, he created his own smart card and programmed it to behave exactly like a legitimate San Francisco card— only Grand could set the value to whatever he wanted.

In the middle of this research, a Chicago TV station asked Grand to comment on parking meters mysteriously failing in the Windy City. In late 2008, the Chicago City Council voted to privatize the city's parking meters. As a result, Chicago Parking Meters, the company that paid $1.2 billion upfront for an exclusive seventy-five-year lease, decided that some neighborhoods would now be assessed more per hour for parking than others.[3] To implement this, the company purchased new electronic parking meters from Cale Parking Systems. However, shortly after the transition, there were reports of vandalism; the *New York Times*'s Chicago bureau reported 579 cases of parking meter vandalism between April and August 2009 alone.[4] Malfunctions were common as well. The new machines charged the wrong rates, failed to issue proof of payment tickets, or refused to accept money. This affected so many machines that the city of Chicago briefly stopped writing parking tickets.[5]

In June 2009, 250 of the newly installed electronic parking meters within a particular neighborhood malfunctioned.[6] Rumors circulated that vigilantes were hacking these electronic meters to protest the new rates. The neighborhood with the broken units had been assessed a particularly high rate.

Grand believed there wasn't any intentional tampering with the Chicago units. A far simpler explanation existed: Apparently Chicago Parking Meters had neglected to test these particular units to see whether they could handle the higher fares before putting them out on the street.[7] This highlights a fallacy in adopting purely electronic systems: They can fail just as much—if not more—than traditional mechanical systems.

As noted elsewhere, electronic systems are no more secure than mechanical systems unless they have additional layers of security— something Grand has found these meters, and most other gadgets, often do not have. Thus, the transition from purely mechanical to hybrid mechanical and electronic parking meters has not gone smoothly.

The first mechanical parking meter was installed in Oklahoma City, Oklahoma, on July 16, 1935.[8] At that time, it was brilliant. A coin inserted in the appropriately sized slot rolled through a channel until it hit a button that recorded within the unit the value of the coin. A timer then marked down the number of minutes allotted to that value of coin until another coin was inserted. For sixty years, parking meter units on city streets worldwide worked in this way.

In 2001, the *New York Post* reported that 7,000 Canadian-made J. J. MacKay hybrid electronic meters in the city were vulnerable to attack using a simple universal TV remote.[9] These MacKay Guardian models used an infrared sensor in the upper left of the LCD panel to allow routine programming. Thus, if anyone obtained a certain model of universal TV remote and held a certain button on that remote for several seconds, the parking meter would reset the counter to zero—meaning fare-paying citizens might find themselves with parking tickets later.

The New York City remote-control vulnerability was apparently well known within the greater parking fare community; yet, it took public disclosure to force the manufacturer to update the software on those units. MacKay, a company that has installed electronic meters in Florida, Massachusetts, New York, Canada, Hong Kong, and elsewhere around the world, made the changes within a few days.

Today, parking meters are taken for granted, particularly by those who live in big cities. We no longer "see" parking meters, and that is a problem since the parking fare business is a $28 billion a year global industry.[10] Despite that large figure, most cities still lose money on parking meters. The classic scenario is coin collectors' skimming coins off the top. There's also the cost of paying the coin collectors to walk from unit to unit. As a result, cities have switched to electronic parking meters that only take electronic cash and can be monitored remotely. As we'll soon see, however, these new units are not without their faults.

For example, some electronic parking meters only take credit cards and therefore need to be interactive and wirelessly connected to a card-processing network. "The credit card based meters verify your credit card in real time," said Grand, "so they connect up to the server through the cell phone network. What's been a problem with credit card meters recently is most people just want to put in money and walk away. They don't want to have to wait for the transaction to be approved before the meter actually grants them the money. There are also some scary unanswered questions—for instance, Is my credit card number stored on the parking meter? or Can my credit card information be plucked from the air by someone monitoring the wireless communications?"

Grand admitted he hadn't used his smart card attack to full advantage: With further research, he could have modified the audit logs, for example. He also could have cleared the coin count. "Say I'm targeting you. I don't like you, and I follow you around," he said. "Or say you cut me off and steal my parking space. Now you go put in your money and walk away, thinking you are okay. I could go and clear the meter. That's a denial of service. Then you'll get a ticket. And now you have to pay it."

In the summer of 2009, Grand took his concerns to San Francisco's Municipal Transit Authority (SFMTA), which listened politely and acknowledged that the San Francisco system had such flaws. But at the end of the day, the authority told him it wasn't interested in defending against high-tech attacks. Grand said SFMTA was more concerned about people who simply didn't pay at all. "The whole of what San Francisco was doing with its audit logs," he said, "was not for fraud detection but to see meter usage." San Francisco wanted to see what meters got the most usage and then to follow in Chicago's footsteps, hiking the rates in certain high-use neighborhoods.

Grand was not invited to the meeting that SFMTA had with vendor MacKay about the problems he found. And MacKay, like other manufacturers mentioned in this book, has never commented publicly on Grand's research, despite widespread media coverage.

"Parking infrastructure in general is completely unprotected, and parking meters are taken for granted," said Grand. "Maybe there's some password protection for the meter maid's PDA, but that doesn't help. If you get ahold of a PDA and you can analyze that—then game over. I could easily obtain smart cards, and I was curious how the system worked, so I chose that method of attack instead. The attack that I did was essentially a replay attack, which is as old as security itself. It shouldn't have been that easy in 2009."

Smart cards are also being used for transit systems, worker-access badges, and even driver's licenses, passports, and credit cards. We progressively entrust more and more personal data to chips that can be tampered with or rewritten with new or modified data and cloned. When we do this, we're losing anonymity and ceding privacy to unknown parties. Perhaps it's time we slowed down and considered the different ways someone could clone or otherwise forge our digital personal information.

1.

Like many millions of people worldwide, I depend on mass transit to get to work in the morning and home again at night. Currently, I'm given a paper card with a magnetic stripe in exchange for a small sum of money. This card, provided it has enough value to cover the costs, allows me to ride considerable distances. I do so anonymously. The value on any one card is limited to around $50, so I can't put several hundred dollars on it. The machinery that operates the turnstiles may have been built thirty or forty years ago, but even at peak transit times, these gates still operate efficiently. With less than $50 at stake, the incentive for fraud remains low, and paper card systems have not been a major target.

But without much public discussion, the gates at each station have gone high-tech with iconized touch pads for smart card use. Rechargeable RFID smart cards—unlike the ones used for parking meters—are the future, or so we are told; therefore, they must be better than paper. But are they?

One argument for electronic transit cards is that electronic systems will catch scofflaws. But as with the electronic parking meter system, someone has to be looking at the log files. Concomitant with layoffs of the human staff at the terminal, there's also a shortage of IT staff looking over the transit system logs. Automated systems could be built with real-time monitoring, but these aren't always in the budget. Thus, smart card transit card systems are likely to lose more money to scofflaws than the "inefficient" paper card systems.

Green issues aside, the paper cards still offer a number of advantages: They are anonymous, date limited, zone limited, and even numbered. The paper cards contain unique physical elements that keep them from being easily copied. The paper card system, used since the 1970s, hasn't been cloned; yet, several teams of researchers have cloned various versions of the new RFID smart cards. The researchers have cracked millions of cards' embedded encryption, which, like that for auto antitheft or GSM mobile phones, is more than twenty years old. And because they are wireless, transit cards can be cloned without an attacker's physically touching the people who own them.

The principle behind smart cards is simple. Not only can a commuter quickly board multiple forms of transportation by waving a single card over a contactless reader, but the system also requires fewer human operators to collect the coinage—everything's done electronically. But such systems are potentially no longer anonymous. Most transit cards encourage you to register and recharge the smart card on-line via credit or debit card or by direct billing to an existing bank account. The linkage of personal information to a specific card makes it possible for the United Kingdom's Military Intelligence, Section 5, to monitor the London Transit Authority (LTA) and track suspected terrorists in real time. The LTA can flag which bus or train a suspect might be riding at any given moment, assuming the account or card hasn't been cloned or otherwise compromised. Alternatives exist. In San Francisco, Clipper cards can be purchased anonymously and recharged at kiosks in the larger stations.[11]

Transit systems use several types of smart cards today, but the Mifare Classic card, made by Philips subsidiary NXP, is used by several million people in several major cities throughout the world, including London and Boston. Research teams, without vendor cooperation, have been able to reverse engineer weak cryptography in the Mifare Classic with a little effort. Imagine what criminal organizations could do.

For its annual Chaos Communication Congress (C3) in 2007, the Chaos Computer Club group invited Karsten Nohl (whom we first met breaking GSM mobile phone encryption in Chapter 3) and Henryk Plotz, a German researcher, to present their research on the underlying cryptography in the Mifare Classic transit cards. Their talk, "Little Security, Despite Obscurity," touched off a wave of controversy around the world.[12] Nohl said later he could crack the Mifare encryption with a laptop computer within thirty seconds.[13]

The Mifare Classic was first introduced in 1994, and among the initial flaws Nohl and Plotz identified was the relatively trivial amount of computation needed to try every possible key. In other words, a fast computer could go through all the available keys in a short amount of time—say, a few seconds. In cryptography this is known as key exhaustion.

Why would transit systems such as London's use a decade-old smart card in its transit system? The Mifare Classic cards cost significantly less than NXP's latest and greatest smart cards. But prior to Nohl and Plotz's research, the cryptographic weaknesses of the Mifare Crypto-1 algorithm was neither suspected nor publicly known. Again, this is security by obscurity.

Not just anyone can break this decades-old algorithm at home. Nohl and Plotz needed a special microscope to slice off and examine layer upon layer of the resin-encased, embedded, electrically erasable, programmable, read-only memory (EEPROM). Each EEPROM holds roughly 10,000 circuits. Only by analyzing each layer were the researchers able to learn the embedded firmware code. Nohl and Plotz didn't have to deconstruct the entire chip; they used advanced computer modeling to predict some of the functions.

Nonetheless, they worked out the Mifare Crypto-1 algorithm. At its heart, they found a sixteen-bit random-number generator. As Grand could with San Francisco's parking meter smart card, the team was able to freeze this generator on the Mifare card so that it produced the same result every time, providing unlimited free travel. They could clone a transit card in London or Boston and, in theory, use it forever.

Once Nohl and Plotz knew the programming, they were able to construct an array of values—a rainbow table like that used for GSM encryption—to help them predict future card values. (It should be noted that while the Mifare Classic contains 1K of EEPROM nonvolatile memory, the Mifare Plus, introduced in 2008, contains up to 128K of EEPROM, which is much harder to crack. Los Angeles Transit, for example, adopted the more secure Mifare Plus system in 2009.[14])

Nohl and Plotz weren't the only researchers looking at the Crypto-1 algorithm. A few months after the C3 demonstration, another group lead by Bart Jacobs at Raboud University in the Netherlands found a different way to crack it. As proof, Jacobs cloned a London transit card (the "Oyster card"), then took his team for a ride on the London transit system—for free.[15] The group even shot several YouTube videos showing what they could do.[16] Informed of the stunt, an LTA official said that the team hadn't proved anything other than that they had cloned one card.[17]

But others took the new attack on Crypto-1 very seriously. In the Netherlands, the Mifare Crypto-1 algorithm also served as the basis for the Dutch government's Rijkspas card, which among other uses, provides worker access to top government buildings. Dutch officials realized the attack could lead to a serious breach of physical security, so the government immediately posted guards outside the affected build-ings until a remedy could be found.[18] In 2009, the Dutch government awarded a new Rijkspas contract to Siemens IT Solutions and Services and Bell ID.[19]

NXP, of course, took the research very seriously: The company filed lawsuits to prevent Jacobs, Nohl, Plotz, and others from publishing the details of their attacks on the Crypto-1 algorithm until a work-around could be implemented. Nohl and Plotz tentatively agreed to delay pub-lication of their results until later in the year, but Jacobs did not. In July 2008, he won in court the right to publish his team's understanding of the algorithm.[20] This raised a legitimate question: Was the world, with several million cards based on the Mifare Crypto-1 algorithm in use, worse off with the code made public?

Security by obscurity, the idea that one is safe because a given sys-tem has such a small footprint in the market, is for all practical pur-poses dead. Apple Computers, for example, can no longer ignore security researchers when they discover a flaw; Apple has been pub-lishing more and more patches. By using a unique system, one may feel more secure than others using a more widely used system or gadget; however, flaws still exist. And, as demonstrated, it's just a matter of time until someone finds them.

As processor technology advances, it's important to know where the flaws in cryptographic systems exist early on. Both the good and bad guys are hammering away at commonly used algorithms and encryp-tion systems, and we want the good guys to win. Replacing an industry standard that is growing weak takes time—years in fact. By sharing the algorithm, not the key, vendors can at least anticipate obsolescence as researchers continually hammer away. (For the record, NXP has sev-eral newer smart cards that offer more robust and secure encryption than the Mifare Classic.)

The Mifare Classic card is enormously popular, and NXP has said it might take years for its current user base to graduate to more secure cards. But the alternative to responsible public disclosure is to have only the bad guys know the cards are weak. Responsible public disclo-sure of flaws creates an environment in which everyone is aware of problems and can work to solve them. But such disclosures are not always received in this spirit.

During the summer of 2008, a third group of researchers, this time from the Massachusetts Institute of Technology (MIT), cracked the Mifare Crypto-1 algorithm. Perhaps hearing about the Raboud University students in the Netherlands, the MIT students decided to see if they could clone a Boston Charlie transit card, also based on the Mifare Classic card. After submitting its paper to the U.S.-based DefCon security conference, the team asked a professor whether it should contact the Massachusetts Bay Transportation Authority (MBTA) for comment.[21] By then the FBI had heard of the trio's proposed presentation at the DefCon conference and advised the MBTA not to allow the paper to be presented. The MBTA argued in court that the students were violating various parts of the Computer Fraud and Abuse Act (18 U.S.C. 1030), passed in 1986 to protect U.S. government computers from hacking, and sought a restraining order.[22] The MBTA alleged that the students were committing fraud and related activity in connection with access devices.[23] A federal judge agreed and blocked the presentation.

This is an unfortunate trend at security conferences of late. At Black Hat DC 2006, RFID vendor HID successfully sued to keep researcher Chris Paget (whom we met breaking GSM encryption in Chapter 3) from revealing flaws in worker-access cards.[24] And Cisco employees physically removed several pages from the bound Black Hat USA 2005 conference material issued to attendees in an attempt to prevent researcher Michael Lynn from presenting on flaws he had discovered in the company's routers.[25] The judge who called for an initial injunction in the MBTA case managed only to block the presentation at DefCon and eventually let the injunction expire. The slides from the talk are now public.[26]

Adding better encryption and authentication, both of which would make the Mifare Classic cards—all smart cards, for that matter—much more secure, would add milliseconds to each transaction and could, during rush hour, critics claim, slow down the admission process. And there's always the possibility that due to bad authentication a card could be rejected, further delaying the queue. Yet, when faced with millions of dollars in lost revenue due to fraud and inefficiency, it would seem prudent for governments to adopt secure systems. Time and time again they do not.

2.

As part of his MBTA transit card cloning presentation at DefCon 16, one of the researchers, Zack Anderson, included a slide showing an MBTA ID card with his photo and the title "Director of Operations, Red Line, 10 Park Plaza." Physical security is not taken seriously enough. A criminal can create a worker's badge just by copying its design. Such simple identity fraud can get a criminal inside a company to commit other kinds of larceny, such as the theft of intellectual property.

In some offices where I've worked, I've flashed my ID card with my finger obscuring either my name or my photo and still gained entrance. The guards only need to see the design of an ID card currently in use by the company. I once entered the world headquarters of one of the largest Internet content providers with a card I knew had expired, and yet I was able to get to my office on the fourth floor, although I was dependent upon others with active badges to work the elevator for me. Purely badge-based physical security systems can be defeated. We feel safer for having an ID badge; yet, we dangle them in plain sight. Anyone can copy them. On a large corporate campus, anyone can pretend to be one of us.

Some ID badges also offer restricted levels of access through RFID. Not only is there a visual design template on the outside (and some offices do not have a template, just a blank card), but there's an RFID chip embedded within. At this same Internet provider, I had once left my paper notebook in the second-floor lounge the previous day. The next morning I attempted to go down to the second floor before regular office hours to retrieve it, only to find the doors there locked and unresponsive to my badge. During normal business hours I had (or at least I thought I had) free movement within the building. When I returned to my fourth-floor office, the phone rang; security wondered what business I had on the second floor before standard office hours. So, in this case the security system worked.

One way around this security is to follow the chief executive officer—or someone like the mail carrier or janitor with access to the entire building—and hope to brush up against his or her smart card worker-access badge to make a digital copy of its access sequence with some kind of RFID reader. The hardware for a surreptitious RFID reader can fit neatly inside a backpack or piece of luggage; these can be fitted at waist height in order to brush up against a worker-access card on a train or subway or even in the street. The readers do not need to make physical contact with the card. Since the CEO might be a little busy, perhaps one should try a janitor or the mail-delivery person instead.

As mentioned in Chapter 2, Adam Laurie has made a career out of taking apart systems. Using open-source tools, he found that many of the vendors in the RFID worker-access card market today use similar coding. Why spend time and money to code something uniquely when you can simply use a recognized standard? Thus, he's been able to scan cards and break them down into two common formats: Q5 from Texas Instruments and Hitag2 from Phillips Semiconductors. Laurie has since written several open-source scripts in Python to read these formats, and they are available on his website.[27]

With a reader and a free download of the script, one can take an RFID access card and reprogram it to make an RFID card match one belonging to someone with high-level access. Laurie did this, and when he told the card vendor about it, the vendor said that since the clone card was of a different form factor (size), it was not a true clone. Laurie respectfully begs to disagree. To an RFID reader, the form factor makes no difference. The card still worked.[28]

Laurie has discussed how a person standing up to two feet away can clone a card with a handful of hardware. The RFID reader can sense when a smart card is available and when one is not because of its signal. This was proven at a security conference in 2009, at which government attendees may have unwittingly gotten tangled in the mix.

In 2009, Laurie and Scottish researcher Zac Franken set up an RFID reader and a video camera at a booth at DefCon 17 in Las Vegas.[29] The two had discussed the idea of doing this at Black Hat DC the previous year, but it didn't work out. For one thing, the venue in Washington,

DC, had been too limited physically. They had planned to position a camera and an RFID reader at a beverage station, and as the conference attendees poured cups of coffee, the reader would scan for any RFID cards on them while the video camera took their picture. The information from the card and the photo would then appear on a screen.[30]

At DefCon, participants attend arguably the harshest security convention in the world and are frequently exposed to such pranks. There is, for example, a "Wall of Sheep" at DefCon, a projected display of unencrypted user names and passwords that have been sniffed out of the 802.11 wireless signals floating invisibly around the conference.

Whereas one should never attend such a conference without first locking down a laptop or mobile device, most people can't prevent their worker-access badges—or any RFID tag for that matter—from broadcasting data without wrapping it securely in tin foil. If the data is presented in an unencrypted form, then others will be able to read it. Perhaps the operative question should be, Why even bring a worker-access badge to a hacker conference?

This topic came up during the annual DefCon panel known as "Meet the Feds," at which federal agents talk about their experiences with cybercriminals. Toward the end of the discussion, someone from the audience mentioned the RFID experiment down the hall. Some federal agents on the panel became concerned and later asked Laurie and Franken to cease their operations—which they did.[31]

Government IDs, if cloned, could grant a cybercriminal access to highly sensitive facilities and computer systems. That was not the intent of the experiment. Afterward, Laurie briefed the feds on what he had done and how easy it was for someone else to do, someone with less honorable intentions. The media card Laurie used to store the data—presumably after only four or five people were scanned—was destroyed.[32]

RFID-enabled worker-access badges can be useful when layered with other security, such as a PIN. Your access card can be cloned or otherwise defeated.[33] What if you were also asked to enter a four-digit code on a keypad near the entrance? That extra layer of authentication would slow down entry, making the badge system much more secure.

Anonymity wouldn't be necessary if you were an employee or even a guest at a private company.

Taking this scenario a step further, a PIN-based system should also be used for exiting. When you leave your office building, most systems currently do not record the event. Tracking employees sounds creepy; think of what human resources could do with this information. It could, however, be very beneficial in an emergency to know how many people are inside the office building. Guests could also be given temporary, disposable RFID-embedded tags like those on common grocery store items.

3.

If you have a dog or cat, chances are the pet has an RFID tag inserted just under the skin. The tags, the size of a grain of rice, are implanted with a hypodermic needle and usually contain a code of up to sixteen digits. Veterinarians can use this code to identify stray animals or pull up pets' medical history. This technology has been used with many different kinds of animals—cats, dogs, pigs, and cows.

But there are human advantages as well. These same VeriChip tags, made by a subsidiary of Applied Digital, have been implanted in some human military personnel. One can quickly identify a soldier with a scan of the wrist. Moreover, access to different parts of a base or project can be granted or denied based on the embedded RFID tag. There are also commercial applications.

Along the Spanish coast, owners of the Baja Beach Club in Barcelona had the idea to do this with its regular bikini-clad guests to make ordering drinks more convenient.[34] Since most people loathe taking their wallets to the beach, where they can easily be lost or stolen, why not inject tiny RFID tags under the skin of high-end beach goers or VIPs? Then, whenever they want to buy a drink, food, or any other service, they need only present a wrist to an RFID reader to have their account debited.

RFID tags have also been used for tracking medical patients in hospitals; however, despite interest from Microsoft, this has been a flop, with fewer people implanted than originally forecast.[35]

In a report to the American Medical Association (AMA), a special committee warned, "At this time, the security of RFID devices has not been fully established. Physicians, therefore, cannot assure patients that the personal information contained on RFID tags will be appropriately protected."[36]

The AMA's Council for Ethical and Judicial Affairs said using RFID tags in humans for medical purposes may improve patient safety by keeping track of patients and their medicines, but the tags may also pose some physical risks, such as infection or relocation within the body. The council suggests that the tags be constructed of materials that promote tissue growth so that they will not move around. The council also mentions that the tags may interfere with electrosurgical devices and defibrillators. Another group found that the chips had promoted cancer in animals.[37]

The tags do not contain personal or even medical information, only a sixteen-digit identifying number. This number links to a more detailed database containing sensitive, personally identifiable data. Thus, someone who cloned a VeriChip could only obtain services as someone else (say, a drink at the Baja Beach Club). But could someone clone an RFID tag?

In 2006, journalist Analee Newitz had her implanted VeriChip read and cloned by researcher Jonathan Westhaus before a live audience at the HOPE Number Six conference in New York. Afterward, Reuters quoted VeriChip spokesman John Procter as saying, "We can't verify what they may or may not have done"; he added, "We haven't seen any first-hand evidence other than what's been reported in the media." Procter concluded by saying, "It's very difficult to steal a VeriChip.... It's much more secure than anything you'd carry around in your wallet."[38]

Westhaus disagreed. "The VeriChip is a repurposed dog tag," he said. "There is no reason (counterfeit housepets?) why it would have been designed with any security features, and in fact it was not."[39]

So far, California and Washington State have both passed laws banning RFID cloning.[40] Unfortunately, California's law stops short of requiring encryption on RFID tags, which would virtually eliminate cloning. Further, the California law could penalize researchers studying the security of RFID tags.

There's a basic authentication problem with this form of RFID implementation. Once you've inserted a tag under your skin, it's there unless you surgically remove it (which can be done). But just as the tag was originally written with data, most tags are left unlocked, meaning someone else can write new data.

Adam Laurie has voluntarily had a VeriChip embedded in his wrist, and he knows full well what can be done with it. In a live demonstration, Laurie used a custom Python script to create a new tag identification for himself.[41] Afterward, if Laurie were scanned by an RFID reader, the veterinary code he used would say he was a dog or a cat.

This writing and rewriting can be used in a number of different scenarios. Say you want to commit a crime, only you want someone else to be at the scene, not you. You can steal someone else's sixteen-digit RFID tag information and use it to overwrite your own. You can become someone you met randomly at the watercooler at work, with access to whatever part of the office building that person has. Maybe the crime you want to commit includes some common practice, such as swiping a wrist to board a train or a bus in London or to enter a closed-off part of an office building. Criminal investigators looking at the logs will not see you but another, innocent person.

Implanted tags may be fine if you want to make a purchase at the beach, but what happens when you go home? Or somewhere you don't want to be tracked? Since an implanted tag can't be turned off, a tag on a card might make more sense.

In public presentations, Laurie has asked if anyone in the audience has a contactless credit card. These have an RFID tag embedded within a layer of resin in the plastic. Using his reader and without the volunteer taking the card out of his or her wallet, Laurie was able to display its contents on the presentation screen—the person's name and the card's account number and expiration date.[42]

After one demonstration, a representative from American Express, whose card had been read, qualified the information, saying that the account number displayed on screen was an alias that could not be used for online purchase. "ExpressPay has multiple security mechanisms," said the representative. "As the payment host, American Express would not verify/authorize an online transaction using just the alias account number. There are several other security mechanisms that would be required in order for payment authorization to take place."[43]

Nonetheless, demonstrations like Laurie's illustrate the potential misuse of RFID technology in the near future. Without handling an RFID-enabled credit card, a thief could sniff its contents just in passing. And the same is true for RFID chips embedded in the human body, worker-access badges, some public transit cards, and even passports.

Currently, more than forty countries use RFID chips in their passports; the United States uses EPC Gen 2 chips.[44] The United Kingdom's Home Office insists on its website that in order for anyone other than a customs official to read one of these new RFID passports, that person would need to have a specialized reader as well as the unique code to unlock the information inside.[45]

London's *Daily Mail* decided to test this statement.[46] The newspaper asked a woman in North London to apply for a passport "rush" so that she could pick it up the same day (this is an important detail). Without examining the new passport, the *Daily Mail* then sealed it in an envelope and gave it directly to Adam Laurie, who had said he could create his own inexpensive e-passport reader, refuting the Home Office.

On the surface, e-passports look similar to the nonelectronic version, except beneath the printed page displaying your photograph and personal information, there's a chip with a wire curled around it, much like the RFID tags on products found at the grocery store or pharmacy. Like those chips, the passport RFID tag is also passive, meaning it contains no battery. In order to be active, it must be energized by a special reader synchronized to the frequency of the tag.

The United Kingdom's Home Office and even the U.S. State Department claim the tags can only be read from a short distance.[47] However, independent studies have shown a reader located within thirty feet can activate the chip—and therefore broadcast its three data files. One file is an electronic version of the passport data. The second is a high-resolution image file of the individual's passport photo. The third file checks to see whether the first two files have been altered.

In addition to the inclusion of a chip, the other important change with e-passports is a printed machine-readable zone (MRZ) strip along the bottom containing numbers and letters. When manually swiped against a bar code reader, the twenty-four-digit code within the MRZ allows the RFID reader to decipher the files within the electronic passport. In other words, the encryption key needed to unlock the passport is printed on the passport itself. This design is part of the International Civil Aviation Organization's RFID passport standard so that poorer countries can adopt the e-passport sooner rather than later (high-powered encryption is still regulated by the U.S. government and cannot be exported to all countries).[48]

So, armed with a copy of the British RFID passport standard, Laurie managed to access the passport's three files. He did so by reading the RFID broadcast, then figuring out the weak encryption needed to unlock the whole passport. According to the international specification for e-passports, in order to guess the MRZ's twenty-four-digit code Laurie needed to know the date of the passport's expiration (usually ten years after the date of issue) as well as the owner's date of birth. Since Laurie knew the passport had been issued that day (due to the rush delivery), he was able to add ten years to guess the expiration date. Using the name on the package and public resources such as social networks like Facebook on the Internet, he guessed the owner's date of birth. With these two pieces of information, he could create the passport's twenty-four-digit code. Without even opening the package, Laurie now had the key to unlock the passport's files. In total, it took Laurie just under four hours to read and successfully copy, or clone, the e-passport.

The ability to clone an e-passport without physically handling it poses grave security problems. For example, a human trafficker may sniff legitimate passports in an airport and clone one in order to smuggle in illegal immigrants. John Hering and Kevin Mahaffey of Flexilis (now Lookout) have taken the act of surreptitiously reading an RFID-enabled passport one step further. If you can see the data inside an e-passport, they reasoned, you can also learn specific details about the individual carrying it, such as his or her nation of origin. In a dramatic demonstration in 2006, Flexilis produced a video showing a mannequin dragged in front of a garbage can; as it passed, the garbage can detonated with an explosion. The trigger? An RFID reader located inside the can read the passport data and trigger the explosion when, for example, the third American passed.[49]

Hering and Mahaffey do have a solution. They say the read zone of the passport needs to be made unidirectional. This can be done with a conductive inlay on both the front and the back of the passport and a special tag. This would prevent the tag from being read when the passport was left partially open.[50]

Other anticloning proposals exist. The U.S. Department of State decided to use EPC Gen 2 RFID tags in U.S. passports because of the need for quick passenger processing at the borders, the lack of personal information (it's just a unique identification number), and the fact that the United States includes a protective sleeve. Missing, however, is a tag identifier (TID), a factory-programmed tag-specific serial number, that prevents someone from reading and cloning the passport ID onto another RFID tag. As noted in a white paper from the University of Washington and RSA Labs, the TID feature hasn't been deployed in the U.S. passports, one of which the researchers managed to clone.[51]

TIDs have also been proposed for RFID tags in U.S. driver's licenses, but so far they have not been used either. Counterfeit driver's licenses are an age-old problem. Using a razor and a photograph to alter the data on a valid license, teens try to make fake licenses to obtain alcohol or cigarettes, among other things. Most of the changes to the conventional driver's license involve high technology.[52] States investing in new licenses display holograms, multiple photographs, and even images that can only be seen under special light.

Some states are using microprinting to publish unique texts within the standard name, date of birth, and driver's license number information. For example, in Oklahoma, the state has microprinted the lyrics from the song "Oklahoma" on its licenses.[53] To make it harder to reproduce, certain words from the song have been omitted. Another change is that in most states today you can no longer wait in line at the

Department of Motor Vehicles (DMV) to receive your license; it will arrive one week later dispatched from a central office. This change was necessary because so many DMV offices had experienced theft, with blank driver's licenses being sold on the black market.[54]

The most radical upgrade, perhaps, will be a national requirement to include an EPC Gen 2 RFID tag on all state-issued U.S. driver's licenses. Originally, part of the USA PATRIOT Act, the Real ID Act called for the use of RFID tags in extended driver's licenses (EDLs) by 2010. Several states, however, chose not to comply. In 2010, the secretary of the Department of Homeland Security proposed an EDL alternative called PASS ID, which requires only a "common machine-readable technology," although this is generally thought to mean inclusion of the EPC Gen 2 RFID smart tag.[55] Another interesting fact from the University of Washington and RSA Labs study is that the researchers were able to read an EDL within a wallet inside a pants pocket from a distance of two meters; in the future, this might give you pause when walking in a crowd.[56]

The underlying problem with driver's licenses, however, doesn't even have to do with the RFID technology. The process of applying for a driver's license is too easily subverted: It's too easy to fake the original birth certificate necessary to obtain a U.S. driver's license. More than 10,000 state and local agencies are authorized to issue birth certificates. Given that diversity, it's almost impossible for individual DMV offices to verify that a certificate presented for authentication has not been forged. Given this bureaucracy, it is also possible to obtain another person's birth certificate from any state if you know enough information to prove that you are that person or his or her legal guardian. Social networking, such as Facebook and other sites that seem to require that you make this information public, have made that possible.

4.

It's not just credit cards, passports, and driver's licenses that we have to worry about. In 2010, retail giant Walmart announced it would put EPC Gen 2 RFID tags on some of its clothing.[57] The RFID chips will be put on removable tags and not sewn into the clothes. Consumers can cut off the tags at home once they've purchased the clothes. From an inventory standpoint, this makes sense. By scanning the shelves, Walmart employees can quickly determine which size of jeans is missing from the display and restock accordingly.[58] Pairing the inventory scan with the scans at the register where items are purchased should reveal which of the 3,750 U.S. Walmart stores have the most "shrinkage" (retail-speak for theft).

Stores like Walgreens and Rite Aid have used RFID tags for years to track pallets of consumables in the warehouse and high-theft items like razor blades and deodorant in the store itself. The move to put RFID tags in clothing is, however, a bit more personal.

Walmart's mass adoption of EPC Gen 2 RFID tags will reduce their price, making the RFID tags less expensive and therefore leading to other uses. To get an idea of what's possible, the Future Store, a supermarket in Rhineberg, Germany, has allowed vendors for several years to RFID-tag items in the store. There, everything from milk to frozen dinners is tagged so that the consumer needn't pass through a check-out lane with a human clerk at the end: The shopping cart reads the tags on the individual items as they are added, electronically tallying the purchase total and charging a registered credit card when the person leaves the store.[59]

An amusement park in the United Kingdom, Alton Towers, uses RFID bracelets in a unique way. Strictly voluntarily, guests wear the bracelets for a personalized video at the end of the day. RFID readers trigger cameras along the rides the guest takes. This footage, from about eight rides, is then intercut with existing professional footage of the park. When the bracelet is returned, the guest can then view the video and decide whether to purchase a DVD as a souvenir.[60] But could such widespread use lower our threshold for privacy loss?

Industry supporters such as Mark Roberti of *RFID Journal* say no. Roberti has taken to task RFID critics who say that RFID tags can be read at distances greater than thirty feet or that personal identity information might be leaked through RFID. He often states, correctly, that no personal information is stored on the RFID tag itself.[61]

Most of the studies showing the benefits of RFID tags in retail stores, however, have come out of the Sam M. Walton College of Business RFID Research Center at the University of Arkansas.[62] And many of the studies have been sponsored by Arkansas-based Walmart and Tyson Foods and Texas-based JCPenney.[63] One paper produced by the University of Arkansas RFID Research Center seeks to dispute common misconceptions about RFID, including scenarios in which a consumer's identity is revealed. For example,

> An electromagnetic field is needed to power a passive tag to allow it to respond to the reader. This field typically only reaches about 10 to 30 feet. Therefore, unless a tag enters this field, the reader has no idea the tag exists. When a tag enters the field, it can be "tracked" (i.e., the reader knows where the tag is because it is within the read zone) and once it leaves the field, one would know where the tag has been (i.e., traced). Outside of the read zones, however, the tag does not emit a signal nor can a reader locate the tag—a passive tag can only be recognized when it is within the electromagnetic field. Therefore, *continuous* tracking of people/objects anywhere in the world would require millions of readers and antennae located in very close proximity to produce the necessary overlapping electromagnetic fields. Even on a smaller scale, to continuously track a box within a 20 acre warehouse would take thousands of readers and antennae—a situation that is simply not economically justified.[64]

But continuous tracking of consumers is exactly what IBM described in a 2006 patent issue titled "Identification and Tracking of Persons Using RFID-Tagged Items in Store Environments."[65] Walmart hasn't said it is locating RFID readers within the store itself, only in the hands of clerks taking inventories. Yet, IBM owns a patent on an RFID system that can start collecting such data every time a customer enters the store.

To refute the RFID distance limitation, researcher Chris Paget demonstrated how he could construct a fairly inexpensive system that would read a generic EPC Gen 2 tag from a distance of 217 feet—far greater than the mere 30 feet generally advertised for RFID.[66] Paget has shown that RFID is part of the radio section of the electromagnetic spectrum. The RFID reader behaves like a radio transmitter, while the RFID tags reflect transmissions back to the reader. A high-power reader beams out radio frequency energy, which in turn energizes, or turns on, the passive RFID tag. By determining how long it takes to reflect back this energy, it is possible to calculate both the direction and range of the item.

If considered in this way, an RFID reader's reach could be extended by connecting it to a yagi antenna (an antenna like the ones used for television reception). A yagi antenna would focus the signals rather than dispersing them, as would be the case with an omnidirectional antenna. This focus would amplify the RFID reader's signal considerably. Paget says any Gen 2 reader has the ability to read any Gen 2 tag. He's talking about store-tagged items as well as enhanced driver's licenses, passports, and border-crossing cards. Thus, if you could go around with a yagi antenna in the backseat of your car, you could read these EPC Gen 2 tags from distances greater than thirty feet.

EPC Gen 2 tags operate in the 902 to 928 MHz range, which the Federal Communications Commission has designated as the industrial, scientific, and medical (ISM) range. It also falls within the ham radio operator's amateur range, although hams typically don't use the 902 to 928 range because of all the potential noise from ISM gadgets. Paget found that, operating under a ham amateur radio operator license, he could use the principles of radio frequencies and decibels to increase his range significantly. As a ham radio device, the RFID tag reader must follow the amateur radio license requirements by, among other things, remaining restricted to 1,500 watts and identifying itself every ten minutes and at the end of every transmission.[67]

Paget notes there are physical limits to how far such an antenna can read. Limitations include other ISM stations, cross talk with other Gen 2 RFID tags, clutter from reflections caused by non–Gen 2 objects, signals that bounce along the ground, and the general curvature of the Earth. Given those constraints, Paget was able to read an EPC Gen 2 tag from a distance of 217 feet. Theoretically, he could read a tag from five hundred feet. The equipment he used cost about $1,000.[68]

As Paget explained in a video interview with the United Kingdom's *Register*, "Potentially you have everyone walking around with driver's licenses in their pockets that can be tracked. It does facilitate very long-range and widespread tracking."[69] Thus, marketers and others can once again learn the biometrics of those passing within thirty feet of a reader and target advertisements according to the zip code, height, weight, age, and gender information contained on the license. It sounds very much like that scene in Steven Spielberg's *Minority Report* (2002), where anonymity is gone and ads know and incorporate your name as you pass.

5.

While RFID tags lack personal information, containing mostly sixteen-digit codes, when taken together the individual tags can produce a unique proxy and thus introduce a new kind of privacy loss. For example, perhaps a customer returns to a store with an RFID chip embedded in his shoe, another sewn into his jeans, and yet another in his baseball cap. Individually the three tags would not be enough to identify the person, but collectively they could. These three tags—just serial numbers—would allow an observer to differentiate one customer from another with a different RFID in her shoe and yet another with an RFID only in her purse. These serial number proxies can also be associated with new purchases made at a store.

It is true that we have this technology today. When you use a store loyalty card at the cash register, the store might associate your card number with a database and print coupons for items you are likely to buy in the future, based on your past purchasing patterns. The difference with the use of RFID is that the store might surreptitiously read your RFID tags and, upon recognizing a returning proxy already in its database, direct you to discounted items you've purchased in the past or to new items you might like based on past purchases. Some customers might like this preferential treatment. Others will not.

EPC Gen 2 RFID tags cannot be turned off. They're energized by the reader, and as we've seen, it doesn't have to be a specialized reader.

Given that RFID tags can be read from a distance of several hundred feet, Paget describes a scenario in which a thief first surveys the serial numbers in a store looking for RFID-tagged items of high value—such as expensive wristwatches—then waits out by the curb, watching for anyone leaving the store with an RFID tag containing those particular serial numbers. The thief might assume the individual holding that product is wealthy, follow him or her home, and later commit a robbery.[70]

How would the thief know when the house was empty? He or she could drive up and down and record all the RFID tags on the block. Some RFIDs would be from driver's licenses. A few days later, the thief might return and notice that the driver's license RFID tag for the house he wanted to rob was missing, suggesting the person was away.

In a more extreme example, Paget has suggested that law enforcement could also beam an RFID reader from a police helicopter high over a neighborhood, allowing officers to identify a stolen item by its serial number or to find a person via his or her driver's license. This could create some interesting Fourth Amendment issues.[71]

Paget notes that surreptitious reads like this could be accomplished with American driver's licenses using the PASS ID system.[72] While this may sound theoretical, Paget tested his theory in 2010 with NEXUS ID cards.[73] NEXUS ID cards are credit card–sized RFID-enabled minipassports. In theory NEXUS can speed up border crossings by broadcasting ahead to customs agents a unique code linked by a database to the next person in line. On one afternoon, while driving in downtown San Francisco, Paget passively recorded several unique NEXUS identification numbers. While their identities weren't revealed, Paget could nonetheless track these people as they continued to move about the city.

Remember that experiment at DefCon in which RFID cards were surreptitiously scanned? With the PASS ID driver's license or NEXUS ID card, you can do the same thing at your local coffee shop. With a laptop and an RFID reader, when a customer reaches over to put cream in her hazelnut coffee, if you are within a few feet, you should be able to pick up her unique driver's license code. The patron will leave having no idea that her personal driver's license data has been swiped but upon return might see a display that says, "Hey, hazelnut coffee person, welcome back!"

Simply put, Paget would like the United States to scrap the whole Western Hemisphere Travel Initiative, a product of the Intelligence Reform and Terrorism Prevention Act of 2004 that requires documents showing nationality for all individuals entering the United States.[74] Documentation being issued for the Western Hemisphere Travel Initiative includes the U.S. e-passport, the NEXUS, FAST, and SENTRI border-crossing cards, and PASS ID driver's licenses.

"I believe that RFID is very unsuitable for tagging people," Paget told the United Kingdom's *Register*. "I don't believe we should have any type of identity documents with RFID tags in them."[75]

As we'll see in the next chapter, even without RFID, we're still giving out a lot of personal information—and in ways we probably haven't considered before.

CHAPTER SIX

The Myth of Fingerprints

A total of 191 people were killed, and 1,800 were injured, some seriously, when ten bombs detonated inside four trains during the busy morning rush hour near Madrid on Thursday, March 11, 2004.[1] The international media considered the event Spain's version of 9/11, although no known link to Al Qaeda was ever established. Amid the debris were at least three bags of unexploded chemicals. One of the bags yielded fingerprints that the FBI initially linked to Brandon Mayfield of Portland, Oregon.[2]

Mayfield, who grew up in Halstead, Kansas, had served in the U.S. military from 1985 to 1989, so his fingerprints were already in federal databases. When the Spanish authorities asked the United States for help in the wake of the bombings, the FBI put the prints found on the bag of explosives through their databases and came up with Mayfield. Six weeks after the bombing, he was arrested at his law office.

According to the indictment against Mayfield, the FBI had verified the match. This clashed with the fact that his family could testify that Mayfield had not been out of the country for the last eleven years. Just one year earlier, Mayfield had defended one of the Portland Seven, a group of American Muslims convicted of attempting to aid the Taliban in Afghanistan after 9/11. He was also a practicing Muslim himself. But the crux of the FBI's case against him was the fingerprint it had matched in its labs.

When the FBI sent Mayfield's prints to Spain, officials there reportedly told the FBI they had other suspects.[3] More than one year later, the FBI would issue a press release stating that "the May 2004 arrest of Brandon Mayfield was based on an extremely unusual confluence of events."[4] Mayfield's prints were one set of twenty identified as possible matches for the prints supplied by the Spanish government.

Ultimately, Hicham Ahmidan, a Moroccan national, was identified as the one leaving fingerprints behind at the scene of the bombing, and in late 2008 a Spanish court sentenced him to ten years in prison for his involvement.[5]

Back in the United States, after seventeen days in U.S. custody, Mayfield was released. He later sued the federal government for violating his Fourth Amendment protections against illegal search and seizure and received an award of $2 million. Additionally, the courts overturned several parts of the USA PATRIOT Act that had made it all too easy for U.S. officials to arrest Mayfield. The government is appealing that decision to the U.S. Supreme Court.

No scientific study has ever shown conclusively that two sets of human fingerprints can be matched with total certainty. Despite the use of fingerprinting by law enforcement and others for over a century, as well as the method's application in security systems protecting buildings and luxury cars, one-to-one fingerprint comparisons are subject to considerable interpretation and may not even be accurate. Yet, we invest millions of dollars in fingerprint technology, and sometimes we use fingerprint matching to send the wrong people to jail.

How accurate can prints from fingers pressed in ink and smudged on a piece of cardboard be? How accurate are impressions of delicate ridges and swirls lifted from a canvas bag? Not only is the fingerprint comparison methodology flawed, but fingerprints themselves aren't eternal. The human hand is the most frequently damaged part of the body, and skin on the fingertip naturally breaks down over time. Consider a construction worker whose fingers might get cut multiple times, abraded, or, worse, severed. Biometric systems that rely on fingerprints include no provisions to handle natural and unnatural changes to the finger over time.

1.

The Mayfield story typifies what Ian O. Angell and other experts say are problems with biometrics in general. Angell, a professor of information systems in the Department of Management at the London School of Economics, told me the problem is one of "residual category," that is, what is left over after a classification is made.[6] Angell explained that in designing an information system—say, a biometric system and, specifically, a fingerprint scanner—software developers categorize different properties and rank them as unique data. What's left over is the residual category. Fingerprints contain unique ridges and whorl patterns said to be unique to a very small number of people. But fingerprint analysis typically does not compare every aspect of a person's finger. Many fingerprint systems only compare a few significant points of reference and ignore the rest. Information-security analyst Joshua Marpet agrees: "We've been telling people that fingerprints are a problem for years."[7]

In the same year as Mayfield's arrest, the British magazine *New Scientist* ran an analysis of a U.S. Department of Justice (DoJ) study claiming a remarkable 2.5 billion points of comparison using 50,000 preexisting fingerprint images. "Contrary to what is generally thought," *New Scientist* authors James Randerson and Andy Coghlan concluded, "there is little scientific basis for assuming that any two supposedly identical fingerprints unequivocally come from the same person."[8] The authors noted that the DoJ study, of course, used too small a sample and that they had found flaws in the DoJ methodology itself. Randerson and Coghlan wrote, "Critics point out that it is hardly surprising that a specific image should turn out to be more like itself than 49,999 other images."[9]

The author of the study, Steven Meagher of the FBI Latent Fingerprint Section, Quantico, Virginia, responded to *New Scientist* in a letter a few weeks later. He clarified that his wasn't a study of error rate or identification but one designed to show how each fingerprint differs from every other one. Meagher admitted that "this technology is not 100% accurate (and requires assumptions and forces limitations that fingerprint experts do not have) but it is the best available to handle large numbers of comparisons quickly."[10]

These academic discussions are troubling in that many court cases worldwide have been decided based on what are now thought to be poor biometric matches.[11] Shortly after Mayfield's experience with the FBI, the U.S. Congress instructed the National Academy of Sciences to conduct a thorough study of the nation's law enforcement forensic capabilities.[12] The report, *Strengthening Forensic Science in the United States: A Path Forward*, released in 2009, was not very encouraging.[13]

In a statement before the U.S. Senate Committee on the Judiciary, the Honorable Harry T. Edwards said the quality of forensic science disciplines was called into question by "the frequent absence of solid scientific research demonstrating the validity of forensic methods, quantifiable measures of reliability and accuracy of forensic analysis, and quantifiable measures of uncertainty in the conclusions of forensic analyses."[14]

The report went on to point out that aside from DNA sampling, which does have an established protocol that can be independently repeated by different practitioners, no other forensic discipline has an established process for matching a piece of evidence to a given suspect. That means that techniques such as fingerprinting and matching bullet casings or trace paint chips—events commonly depicted on network television shows such as *CSI*—are not scientifically defensible and are subject to interpretation and error. The National Academy of Sciences report implies that faulty forensics have put innocent people in jail and left criminals on the loose.

That hasn't stopped others from attaching great value to the ability to match fingerprints. Grocery store chains have begun experimenting with in-store payment systems that use fingerprint-scanning systems to access customers' bank accounts directly. The systems look for only a few fingerprint points to identify you, linking your live print with your bank or credit card account. Some customers like the convenience of not reaching into a purse or wallet to make payment. Instead, they can simply insert a finger into the scanning gadget, and the system attempts to match the scanned image with an image on record. But such systems are easy to compromise.

There are several ways to defeat simple fingerprint scanners. The typical scanner takes a picture every time you press your finger down. Such a gadget only "sees" a flat image of your fingertip. Thus, you could take a photograph of someone else's finger and press that image up against the scanner. You could also, like they do in the movies, have a drink with a friend, then afterwards, using black talcum powder and transparent tape, lift a latent print—called a found print—off the drinking glass and present that flat fingerprint image to the scanner. Some savvy scanners, aware of this trick, now look for moisture. Adam Savage and Jamie Hyneman, hosts of the Discovery Channel's *Myth Busters*, have demonstrated how you can fool such a fingerprint scanner.[15] To create moisture, the *Myth Buster* duo simply licked the fingerprint image before presenting it to the scanner.

Fingerprint scanner vendors say they also look for the roundness of the finger and the pressure applied to the scanner. Back in 2001, security researcher Tsutomu Matsumoto had an answer for that, too, one involving gelatin (like that found in gummy bears and other sweets).[16] Matsumoto etched a latent fingerprint from a drinking glass onto a photosensitive circuit board, then poured gelatin over this high-contrast and very accurate fingerprint reproduction. The fake finger worked 80 percent of the time against different fingerprint-scanning systems.[17]

These methods of defeating common fingerprint scanners sound complicated, and some of them are. But it all depends on what's on the other side. If you want to gain access to a military facility or even just to buy a few hundred dollars' worth of groceries on someone else's tab, then maybe it's worth the effort. Some European countries are using fingerprints for immigration control, and given the questions around the technology, that decision may be premature.[18]

The Chaos Computer Club (CCC), a long-standing German hackers' club, has published various ways to capture fingerprints, transfer them onto a foil surface, and then use the print to confound the biometric readers used by the German government. To raise awareness about the government's growing fingerprint collection—and in an act that may have violated Germany's privacy laws—the CCC published the fingerprint images of then German home secretary Wolfgang Schauble and others in the German government. The CCC said the next time you're asked to give a fingerprint, use Schauble's instead.[19]

2.

What if we don't need an exact fingerprint match? What if we want to design a fingerprint system that prevents another individual from using a ticket or a card but doesn't necessarily identify him or her? In the mid-1990s, the Walt Disney Corporation faced this problem. Its theme parks were experiencing losses from abuses of its annual and season passes, which technology made it easier to counterfeit. What had been in place, bar codes and laminated photos, could now be copied by affordable desktop computer graphics software and low-cost color scanners and printers, creating problems much like those with fake driver's licenses. The already overwhelmed ticket takers at the park could not possibly scrutinize each and every one. Disney needed a convenient way to screen for forgeries and fraudulent use.

In 1996 the Disney theme parks began experimenting with fingerprint readers. The machines scanned only the index finger and put an image of the fingerprint on a magnetic strip in a process much like that used by state departments of motor vehicles (DMV). At first the company did this only for annual and season pass holders, since those tickets held the most value. Almost ten years later, Walt Disney theme parks worldwide instituted a radical new policy: All ticket holders—not just annual and season pass holders—had to be fingerprinted upon entry with a new system.[20] Adult patrons—minors were exempt—had to place their index and middle fingers in a gadget that recorded on the ticket not an image of the fingers but data about each finger's unique shape and position relative to its neighbor. On repeat visits, if the two fingers presented matched those recorded on the ticket, then the person would be admitted.

In early 2008, Disney upgraded its fingerprint-scanning system. The new machines now required a live scan of one finger—usually the index—and compared the result to the original record stored on the ticket. As noted elsewhere, although the scanner takes an image of the entire fingerprint, it only records a small set of points. Some point-based fingerprint system vendors encode minutiae, tiny details of a fingerprint, and call the fingerprint converted into a number "you." Instead of the whole fingerprint image, like those seen on the television show *CSI*, only the unique number is recorded on the ticket, not the full image.

In practice, the gate-access process at Disney theme parks is similar to that for boarding a high-tech subway, with an extra step: Insert a ticket, press your finger against a scanner, and if the points of the live finger match the points on the recorded file, the gate is opened, and the ticket is ejected on the other side. Since no personal information other than the numerical points related to fingerprint details is stored on the ticket, the ticket can only tell the ticket taker whether it has been forged or transferred to another party. The system is designed to tell the parks whether the person entering today is the same person who entered yesterday.

Disney has said it keeps its fingerprint system separate from its other computers and purges the data thirty days after the ticket expires. Visitors who object to fingerprint scanning can show a ticket clerk a photo ID instead, although that option is often not well advertised at park entrances. But if the park doesn't store personal information about the ticket holder from the ID, how is checking the ID going to prevent fraudulent use?[21]

Things get interesting when the Disney system fails. If you purchase a group of tickets, only one member of the group needs to be identified. What if the person in your group who recorded a fingerprint is back at the hotel with the flu or returned home early for a work emergency: In theory you won't be able to enter. Park officials may then ask to see some form of ID from one of the remaining group members. Disney is not alone. Six Flags, SeaWorld, Bush Gardens, and Universal Studios have all instituted similar fraud-protection policies at their gates. The theme parks generally claim that the technology reduces wait times and prevents fraudulent use of multiday passes. Sometimes when lines are long at the parks, however, the fingerprint scanners are simply turned off.[22]

As for storing only the specific points of a fingerprint and not its whole image, scientific methods exist for reconstructing the original fingerprint. Researchers have taken the stored points of a fingerprint and used algorithms to calculate the thousands of missing ridges and whorls.[23] Thus, with the right algorithms and a little computational horsepower, one can create a reasonably close facsimile of the original fingerprint. In another approach, Canadian researcher Andy Adler uses what's called a hill-climbing attack, in which digital noise is systematically added to a point-based fingerprint template until enough of the missing data has been re-created.[24]

Both methods require a heavy amount of calculation, but if, as we have seen in Chapter 1, laptops today are fast enough to re-create the antitheft encryption used on cars, the day when everyday laptops can be used to crunch these biometric algorithms is also fast approaching. Rather than scrap existing fingerprint systems entirely, researchers suggest that point-based fingerprint system vendors exclude certain salient minutiae from their templates. But if a thief can reconstruct a fingerprint from a point-based template on a discarded ticket, then use Tsutomu Matsumoto's research to make a gummy bear finger mold, that person could then use the forged fingertip for just about anything. Then where would we be?

3.

Consider the story of K. Kumaran, a Kuala Lumpur accountant and carjacking victim.[25] Leaving work one night in a Kuala Lumpur suburb, Kumaran was knocked down by a gang of car thieves driving a Honda Civic. Kumaran owned a secondhand S-Class Mercedes-Benz with a biometric door lock.[26] The men threatened Kumaran with Malaysian machetes, demanding that he give them his car. He later told Malaysia's *New Straits Times*, "They forced me to put my finger on the panel and then started the car. They bundled me into the back, between the seats and used my tie to blindfold me," he said.[27]

The thieves took Kumaran to an auto chop shop, apparently hoping they could defeat the biometric security. When they could not, they instead cut off the tip of his left index finger. They dropped him off along the roadside, where he was eventually able to find medical help.[28]

It took another crime to draw attention to a newer form of finger biometrics. From 2000 through 2005, Japan experienced a string of impressive ATM robberies. At the time, Japanese consumers had no withdrawal limits on their accounts, and banks didn't guarantee against losses due to fraud. Due to public outcry, the Japanese government shifted responsibility for such losses to the banks.[29] The Japanese banks quickly turned to vein-pattern recognition as a means of authenticating ATM users.[30] Two of the four major Japanese banks aligned with a fingerprint-based vein-pattern recognition system made by Hitachi. The other two aligned with a palm-based vein-pattern recognition system made by Fujitsu.

Vein-pattern recognition uses near-infrared light to look for the oxygen-depleted blood contained in veins; the resulting vascular pattern is considered to be unique for each individual, even identical twins.[31] The deoxidized hemoglobin absorbs the near-infrared light at a certain wavelength and appears as a black pattern in an X-ray-like image. An algorithm then converts the image into a mathematical template. This template is stored on a card or within the network.

When a cardholder requests funds from an ATM, the card is inserted, a PIN is typed, and either the palm or a finger is held over a scanner. This is three-factor authentication: a card (something you have), a PIN (something you know), and your finger or palm (something you are).

This system is also being used in hospitals. I toured one facility in Pleasanton, California, that uses the Fujitsu palm-recognition system for admission. A patient puts his palm on a plastic holder, and light scans the reflected vein pattern. A nurse asks the patient for his birth date. This is two-factor authentication: something you know (birth date) and something you are (palm). As with the tickets at Disneyland, the process here only confirms that your finger or palm matches the information you provided at enrollment. It doesn't really confirm your identity, only that you are who you registered as when you enrolled.

The Japanese ATM system is very effective. The Bank of Tokyo-Mitsubishi, Japan's largest, has had a Fujitsu system in place for several years in 5,000 locations, serving over 1 million users without incident.[32]

Unlike the fingerprint scanner used on Kumaran's car, vein-pattern recognition requires liveliness, or flowing blood. A cadaver would not produce an acceptable vein pattern. So, to a carjacker, a severed finger would be useless on a vein-pattern system. Unfortunately for Kumaran, such systems are not used on cars today.

We're not just protecting expensive automobiles and bank accounts with biometric systems. After 9/11, the Federal Aviation Administration (FAA) and other U.S. government agencies convened a panel to address airline passenger protection and identity verification using biometrics. One member was Gordon Levin, a Disney staff engineer experienced with biometric systems in use at the parks. It seemed a perfect fit. Just as Disney struggled with reducing wait times and hassle for its guests, the FAA sought an efficient means to verify airline travelers—or rather, to exclude the possibility that some attempting to fly were really terrorists. The FAA decided to pass on fingerprint scanners, but the Department of Homeland Security did not; it uses fingerprint scans at the U.S. borders with Mexico and Canada. Besides Levin, several former Disney employees have joined the National Security Agency and the U.S. Department of Homeland Security in the years since 2001.[33]

The FAA ultimately decided to combine fingerprint and iris scans, creating more layers of security. The FAA awarded a Virginia-based company, Verified Identity Pass Inc., with a prevetted airport-security program called Clear. While operational, Clear stored both fingerprints and iris scans on a smart card. As with the tickets at Disneyland, individuals carried a copy of their Clear electronic smart card; when reentering the country, they put their card into a reader, had their iris and index finger scanned, and if the two scans matched what was on the card, the individuals were admitted, bypassing lines at customs. Other perks included cutting to the front of the security line at boarding. In the summer of 2009, the Clear program ended as the result of mounting debt. As of this writing, it remains unclear either what became of the biometric data Verified Identity Pass Inc. had in its possession or who now owns that data.[34]

Despite the failure of Clear, iris-scan comparisons are considered more accurate than fingerprint comparisons. The system is passive: Cameras record the irises of individuals from three or four feet away as they walk through a line. Iris-scanning systems look at the unique position of irides, the colored membranes that control the amount of light reaching the retina. This is also the part of the eye people refer to when they say someone has blue, brown, or hazel eyes. A typical stored digital photograph of a person's iris has about two hundred pixels and contains relatively stable information: Unlike a fingerprint, irises don't abrade or otherwise change, though they can be obscured by cataracts with age.

One commercially deployed iris-recognition algorithm is John Daugman's IrisCode. In a paper, Daugman showed how even genetically identical people (twins) have uncorrelated vein patterns in their irises, making the iris truly unique.[35] Daugman's IrisCode has an unprecedented accuracy rate; not a single false match has ever been reported for this algorithm in more than 200 billion combinations of iris pairs scanned worldwide over the last ten years.

At Heathrow International Airport the Iris Recognition Immigration System records travelers' irises.[36] Stepping inside a booth, a person need only look at a fixed point—a dot that appears to be hovering midair and must be held in the center of one's field of vision to work. Often that entails a few seconds of standing back or moving to the left to make physical adjustments. While the technology itself—the scanning of the retina—appears to work, the degree to which the British and other governments accept that fact does not. At Heathrow, according to security researcher Zac Franken, after the retina scan confirms your identity, a door opens, and you are admitted back into your home country without the need to show your passport.[37] This paves the way for the day when a high-resolution photo of a particular iris can mimic a live human being. When that happens, there is no backup procedure. Perhaps, the British and other governments should first accept the iris scan, then also ask to see your passport.

Retina scanning is even more accurate, although this technology is rarer because it is more intensive, measuring the live blood vessels and vein patterns along the back wall of the eyeball. The veins along the back wall are unique—even among identical twins—and therefore can be mapped and stored digitally as a positive identifier. One system works by having the user center his eye on a fixed mark while a scanner reads a circle around that point. Moving around the circle, for every blood vessel the scanner records a one; for all else, it records a zero. The original file format for this scan was 360 bits, one bit for each angle of a circle, and the resulting pattern, such as 00010011000 . . . , is said to be unique.

Security researcher Joshua Marpet calls retina scanning the gold standard of biometrics: "There is nothing better."

Retina scanning cannot be counterfeited, but other physical limitations exist, such as disease. Nor is the process very user-friendly. Marpet described one system that uses a puff of air to take a scan—most people do not want a puff of air shot at their eyeball. And while the scanner takes its read, it's hard to hold still. One twitch and the scan has to start over. Other factors can influence the results, such as severe astigmatism, aging (as people grow older, their bodies change naturally), and disease (such as macular degeneration).

So far we're talking only one-on-one comparisons. What happens when we use this biometric technology to pick someone out of a crowd?

4.

In 2001, the Super Bowl was held in Tampa, Florida. That year, the Baltimore Ravens, winning a wild card berth at the Raymond James Stadium, managed to defeat the New York Giants 34–7.[38] That Super Bowl is memorable, however, for another reason. As fans entered the stadium, one of four facial-recognition system companies used cameras to photograph each patron.[39] These images were then fed into a federal database and matched with known or wanted criminals. The attendees were not told of this until after the event.

Today, the Super Bowl is considered a national-security special event, allowing the U.S. Secret Service wide investigative leeway in the wake of 9/11. But in January 2001, nine months before 9/11, the government had no such latitude. The experiment at the Raymond James Stadium was conducted to determine what kind of security might be used at the 2002 Salt Lake City Olympics. After the Super Bowl, the American Civil Liberties Union in Tampa immediately filed suit, arguing the facial-recognition scans violated Fourth Amendment protections against illegal search and seizure (this was a few years *before* Mayfield's arrest).

Legal issues aside, the experiment was deemed a technical failure. Not many matches were made. Facial recognition using crowds was not repeated at the following year's Super Bowl; nor was it used at the 2002 Olympics.

If fingerprint systems store too little data, and vein-pattern, iris, and retina scans are sometimes inconvenient, the face would seem sufficiently complex to be unique, and imaging it should be fairly nonintrusive. How many photographs have we all appeared in?

But Ian Angell says that facial-recognition systems are especially problematic for biometrics. Certain ethnicities do not photograph well enough for facial-recognition software to distinguish bone structure. He showed me how a simple facial tic can make the system fail. "You aren't allowed to smile," he said. No shadows. "They're forcing you to conform to their photographic system."[40]

Some DMV offices in the United States are now requiring licensees to stare blandly at the camera. This no-smile policy is to allow compatibility with facial-recognition systems, which these states may purchase in the future. How practical is that?

When walking down the street, you don't hold your face absolutely still, so how are these systems going to find some random person on the street? Or entering a sports stadium? You need only put on a mustache or maybe a little bit of makeup, and the computer thinks you're not you because it hides your bone structure. Or maybe you've been up all night, and you have big black circles under your eyes. These shadows change the way the software interprets your bone structure. Again, facial recognition doesn't work.

Marpet agrees: "Faces are very complex. If you try and [model] a human face, it takes a lot of work." The human face has forty to fifty muscles, depending how you count.[41] Think of any Pixar-animated movie; the faces aren't very detailed. Just getting a computer to create a realistic walk is resource intensive enough. "If you want to map a human face, it takes a hell of a lot of computing power," said Marpet. "And you can change all that work just by smiling."

The simple act of smiling changes your whole face; it does so by moving your forehead, your ears, your mouth, and even your nose. "What we need," said Marpet, "is a mythical computer system that can map your entire face at rest, then has you go through a range of expressions so it can interpolate between those expressions to whatever expression you may have later on."

Can you imagine such a complete system? Say you start a new job or, worse, need to get a new photo at the DMV. The person behind the camera says, "Relax your face. Now smile. Be pissed off. Be surprised." This would capture all the different faces to be stored as you.

Marpet laughed, "What, am I in acting class here?"

Even if we did submit to a twenty-minute photo session, storing one good high-resolution image takes space, time, and effort. One image may take up 1 GB. Ten photos would take 10 GB. Now multiply that by the number of people within the state. Marpet says the state of California would need a storage facility the size of a building. And it gets expensive, not just for storage but also for the computer resources necessary to interpolate all the expressions we're capable of that the camera didn't record. You might not be smiling when you rob a bank.

Marpet put it simply: "Computers don't see things the way we do. In biometrics it's all about correlating what a computer sees or senses versus the Eyeball Mark One [and the human brain]," he said. "You have two eyes, two ears, a nose, a mouth. What's so hard about that? Well, the problem is that the guy next to you also has two eyes, two ears, a nose, and mouth. Still, what's so hard? Well, the woman next to him also has two eyes, two ears, a nose, and a mouth. What's the difference? It's subtle, and to a computer it may be too subtle."

Even human beings do a poor job of distinguishing one person from another. Before fingerprinting was accepted for law enforcement use over one hundred years ago, there was a system for making sure that people were identified. In the late 1800s French criminologist Alphonse Bertillon started measuring bodily features of habitual criminals, including photographs (the first mug shots). Of Bertillon, Marpet said, "He would bring in the rogues—these are the criminals of the time. He would measure their head; he would measure the length of their index finger and length of the ring finger. He would measure the width of their head across the widest point, the length, the distance from the nose to the mouth, from nose to eyes, how far apart the eyes were. . . . So he was doing cranial and bodily measurements, and this was before fingerprints came around. And that's what they were doing to identify criminals." This was an early form of biometrics, measuring criminals when they were caught and seeing if those measurements matched an existing record. Interestingly, Bertillon initially opposed the use of fingerprints for identification, although he later accepted the practice.[42]

"We are hardwired to try and categorize things," said Marpet. "We try to break things down to numbers. Yet, humans are very complex individuals, but remember I said from my point of view, even using Ian Angell's organizational system, what we're trying to quantize is the residuals. We're trying to quantize the little bits. The pieces that are very, very similar in most people. I have to identify you from you? God. You're asking me to do a level of detail that is very hard. But as a human, do I do that instinctively? We do not."

Marpet, who is white, cites an example of moving to Asia: He said he initially had trouble identifying one person from another. "You will see a sea of people who all look like *x*. In China, a lot of people are from Guangdong Province, which is a significant portion of China, and they have straight black hair, angled eyes—they don't have the round eyes—they have a certain cast of skin, a certain set of features. Only after I'm there for six months can I start to tell Johnny from whomever. But for those first few months, it's difficult. Even humans have to train their eyes to see other humans differently."

As a counterexample, Marpet cites computer versus human recognition. "Say you put two iPhones side by side on the tabletop. A human would be able to say quickly, 'That's mine,' because of the particular wear patterns on the case," he said. "Another difference: One phone has certain apps the other does not. The point is, come back an hour later, and the two iPhones will still have more or less the same differences."

"The problem with human recognition," said Marpet, "is that we're made of flesh, not plastic; flesh moves and plastic does not. So a photograph of us one minute might not match a photograph of us of a minute later."

At Black Hat DC 2010, Marpet presented an illustration showing two faces and dots of light displayed in between them. The points of light indicate the differences between the two images—this is what the computer sees, only the differences and not the faces themselves.[43] First, he showed a pair of images of two different people wearing the same expression. Black and white, male and female, it didn't matter: The differences between them registered by the computer were slight; only a few points of light. He then showed two photos of the same woman, except in one image she wore a neutral expression, and in the other she was very angry. Guess what? The points of difference between her two images greatly outnumbered those in the comparison between two different people. This is an example of how a computer could assume two images of the same woman were photos of two different people. This would yield a false positive result.

It isn't just that our facial musculature and flesh move. Other external conditions can also affect the accuracy of facial recognition. In 2007, the German government conducted one of the first large-scale scientific studies of facial-recognition software by cataloging passengers streaming through a Mainz train station. The study, conducted from October 2006 through January 2007, found that the program could match volunteers only 60 percent of the time during the day and only 10 to 20 percent of the time at night.[44] Why such a difference? The computer program was unable to account for differences in light on the subjects' faces as eye sockets and cheekbones may appear different under artificial light. The study also found a false-positive rate of 0.1 percent. That may not seem high, but given that Mainz sees 23,000 passengers a day, that means twenty-three people might be pulled out of line and subjected to additional questioning by German authorities. While the police might not have a problem questioning twenty-three people each day, the false-positive interviewees might feel otherwise.

Even localized, personal facial-recognition systems—which, like the Disney ticketing system, only identify you as the owner and not someone else—have their faults. In a presentation titled "Your Face Is Not Your Password," a team of Vietnamese researchers, lead by Nguyen Minh Duc from Vietnam's Hanoi University, demonstrated how to defeat facial-recognition systems on certain Lenovo, Asus, and Toshiba laptops using a photograph or drawing. These systems use built-in webcams to see and software to block unidentified users from accessing the laptop's contents. Like fingerprint scanners, they look for points of difference with the recorded image. As we just saw, however, even two different photos of the same person can appear vastly different to a computer. On the other hand, two similar photos of two different people wearing the same expression can look identical to a computer. Thus, the researchers were able to use photographs, even fake pictures of faces—drawings no less—to defeat the biometric authentication on these laptops.[45]

"Say that facial recognition does work 80 percent of the time," Angell said. "That still isn't any good. Maybe 90 or 95 percent . . . but that's about your limit. There's 5 percent that doesn't work." And for every percentage that you're off by, you start producing more Brandon Mayfields.

5.

Some passive facial-recognition systems do work, but only within the strict limits Angell describes. Consistent photographs, such as those obtained from government-issued driver's licenses or passports, work best. A system at the DMV in Illinois, which compares DMV photos over the years with names and addresses, has found more than 5,000 cases of identity fraud from 1997 to 2007—often one person obtaining a second license under an assumed name and address.[46] In 2006, a team at the Registry of Motor Vehicles (RMV) in Massachusetts wanted to see if they could use facial-recognition software on their records to identify false names and addresses, as well as criminals, in their database.

For example, the Massachusetts RMV took a televised photo of Robert Howell, a fugitive wanted in Massachusetts on a rape charge, from Fox's *America's Most Wanted* television show and compared it against the 9 million faces stored in the RMV database. Remarkably, they got a hit—a second name and address used by Howell. They then sent this information out to neighboring states and soon discovered Howell living under this alias in New York State, receiving New York State welfare. Howell was a rare exception.[47]

Each day the Massachusetts RMV processes more than 5,000 new driver's license requests. The facial-recognition algorithms scan over 8,000 facial data points, producing a list of potential duplicates. Then a human being, not a computer, must sort through them. Most are quickly eliminated. Some of the matches are clerical errors—legitimate license updates, for example. Some are fraudulent, with people often wearing the same clothes, even though they are applying under a different name or even in a different state. In one case, among the potential duplicates were 157 sets of twins.

In the absence of standards, states are saying that those getting new driver's licenses should not smile, as it can influence the effort to develop supersecure driver's licenses and foolproof identification cards in the future.[48] Although many DMVs do not currently use the technology, neutral facial expressions are required in Arkansas, Indiana, Nebraska, Nevada, and Virginia.[49] According to the American Association of Motor Vehicle Administrators, like Illinois and Massachusetts, thirty-one states currently look for matches in their state driver's license photos, with three other states considering the technology.

Capturing neutral facial expressions has its downside as well. There's the story of Baroness Joyce Anelay of St. Johns, from the House of Lords. She headed a group that was going to visit the site that makes facial-recognition photo-imaging systems in the United Kingdom.[50]

The company tried time after time to register her photo and, according to Angell, failed. In the end, a man there said, "I'm sorry, but your face is too bland."[51]

The goings-on within individual state driver's license bureaus, however, are small potatoes when compared to the efforts of the U.S. government and its plans for the future. Since 2008, the FBI has been expanding its fingerprint database to include iris patterns, facial-recognition data, tattoos, scars, and perhaps even the particular way a person talks and walks. Called Next Generation Identification, the database will also contain criminal records, the FBI's Terrorist Screening Center's database, and the National Crime Information Center database, which tracks felons, fugitives, and terrorists. By looking at the stored biometric information, the FBI says, it should be able to spot someone living in the United States under an assumed name.

As a test, there already exists a biometric database for every Iraqi citizen who requires access to U.S. military bases. Since 2006 the U.S. military has been taking fingerprints, iris scans, and photographs not to identify these specific Iraqi citizens but to cross-check their data with the database and rule them out as known terrorists.[52] A similar effort is under way in Great Britain, where the FBI is assisting the government in creating a massive biometric database for every UK citizen. Again, the database is intended to prove not that one is a UK citizen but that a given individual is not an illegal immigrant or a terrorist.[53]

Such databases are known to have problems.

Take, for example, Sister McPhee, head of the Catholic Education Service in North America. Shortly after 9/11, the sixty-two-year-old Dominican nun couldn't get on a plane in the United States because every time she went to an airport, she had nothing but grief. "I missed key addresses I was to give," she told a reporter. "I finally got to the point where I always checked my bag, because after I got through the police clearance, then they would put me through special security where they wand you from head to foot all over. They would dump out everything in your bag, then roll it into a ball and hand it back to you."[54]

Angell said an Algerian terrorist, a man in his thirties, had used Sister McPhee's name and thus landed her on the No Fly List. "Here was a seventy-year-old nun, and nobody had the wits to say, 'Hang on. This name is allocated to an Algerian in his thirties, and here I've got a lady in her seventies,'" he said. "They didn't use common sense."[55]

6.

One of the least obtrusive forms of biometrics is voice-pattern recognition. The technology has endured years of bad press because it was marketed as foolproof, after which security researchers found a number of ways to circumvent it. For example, identical twins have the same sinus and throat structures and often sound exactly alike to a machine. Or a registered person might have a cold, so the system rejects the match. Or maybe a voice is recorded and played back in what's called a replay attack.

Voice-pattern recognition works by comparing the sounds made by a particular human throat with a recorded sample. As with fingerprint analysis, only a mathematical representation of that voice is stored. In the early days, this technique was vastly limited by how much data could be collected from a given sample. Today, voice-pattern recognition systems use hundreds of different data points to compare inflection, tone, and other qualities.

Too much background noise remains a problem. For example, such a system, in order to validate my purchase at a noisy store, would need to filter out ambient noise, yet hear my voice clearly. Some systems can already do that, but most still fail.

Another promising new biometric is human gait monitoring. Although facial-recognition systems failed to identify people in crowds at the 2001 Super Bowl, gait-recognition systems work best in crowds. Perhaps you've seen how major athletes train. They practice in front of a camera, which turns key points (say, an elbow, a knee, the head) into data. This data can be analyzed and compared with that from others (say, Andy Roddick versus Andy Murray on the tennis court). Gait-recognition systems are similar. An individual's movements are digitally mapped in dots, and that unique representation is recorded. It doesn't matter if you walk toward a door singly or in a pack with others;

the gait-recognition system will isolate you by the distinct way your various points move. That becomes your signature.

This technology works best over the short term—say, a period of employment at a government facility. Over time, as with fingerprints, aspects of your gait may change in rather significant ways. You might become incapacitated and require the use of a wheelchair or prosthesis, or a progressive disease might set in. Even age changes a variety of aspects of our gait. Sciatica, for example, causes intermittent limping, as does a sprained ankle. Illness, injury, and aging can also change a biometric data point.

"When companies use the tools of a technology to solve a problem," said Angell, "they may or may not succeed, but what is certain is that completely unexpected phenomena happen."[56] That is true with biometrics. Even if everything goes as planned with technology, there are always unintended consequences. As systems attempt to categorize movement in humans, a computer will discard unique clues that a human might not. This is Angell's residual category.

7.

The notion that our technological data-collection techniques are infallible is as incorrect as it is widespread. Popular media created the so-called *CSI* effect, whereby juries now expect crime scene investigators to come in and prove beyond a doubt that a person did this or that. Who cares about the alibi? The reality, however, is very different from what you see on a TV show, as the National Academy of Sciences found. For bullet ballistics, there are no established processes. Tool-mark analysis, which examines the unique marks left on a slug by the barrel of a particular gun, is another area with no industrywide standards. Thus, two different investigations of the same bullet could lead to two different results.

By contrast, DNA sampling has become statistically defensible, using processes that can be replicated by different labs to produce quantifiable and acceptable results within some constraints. "If I wanted a 100 percent absolutely correct system to identify you from me," said Marpet, "I could pull DNA from you, and I pull DNA from me, and I would check the blood against each other. I'm pretty certain we could determine [the difference] between you and me one hundred times out of one hundred times. However, the constraint is, I don't think people are going to agree to have their blood drawn to give DNA every time they go to the airport."

There are several constraints, actually. A simple cheek swab would work; although blood is the best for DNA samples, it's much harder to take blood samples than cheek swabs. Marpet should know. He's a former cop certified to take cheek swabs. Another constraint is privacy. "I'm certainly not going to agree to have my blood stored in the Combined DNA Index System (CODIS) or in the system forever just to make certain it is me or not me."

Is there a way to create a nearly perfect biometric recognition system? Marpet says yes. "Barring any screwup with the specimens, barring any screwup with the mechanics of DNA comparison, other than a failure of the system itself rather than failures of the underlying principles of the system," he said, "DNA should work 100 percent of the time."

For years, the Center for Wrongful Convictions, facilitated by the Northwestern University Law School, has worked to provide DNA evidence for death row cases in which convictions were handed down before DNA evidence existed.[57] In 2003, the group provided its first DNA evidence to exonerate a death row inmate—and later used DNA to lead to the arrest of the correct individual.[58]

Despite the relative accuracy of DNA testing, however, at least one company thinks the methods have failed to keep up with antiforensics technology—technology designed to defeat modern forensic analysis of a crime scene.

In the fall of 2009, the Israeli firm Nucleix released a study finding that it could fabricate blood and saliva samples from one person and make test results identify another individual. The study called into question the level of sophistication of DNA testing, arguing it stops short of conclusively identifying individuals. Indeed, in the O. J. Simpson case, the odds that the blood found on the glove entered into evidence came from anyone but Simpson were 1 in 70 million.[59]

Using genome amplification, Nucleix claims it can use DNA from a hair sample or even from saliva residue on a drinking glass to amplify the DNA in someone else's sample. Someone could take trace DNA from a celebrity and fabricate blood or saliva at a crime scene.

Marpet sees a downside to the Nucleix system. "I can see [their research being] useful for trial lawyers trying to cast doubt on the evidence." This might create short-term gains for the lawyers, but in the long term, DNA analysis is more advanced than other biometrics.

Marpet has been thinking about how to clone biological information. "If you have Dolly the sheep," he said, "can you tell Dolly the sheep from Dolly the sheep?" The answer is yes. Telomeres are regions of repetitive DNA at the ends of our chromosomes, and as we age they unravel. So the telomeres on the original Dolly should be more unraveled than those of the cloned Dolly. "The problem," Marpet said, "is that I'm not taking continuing samples from myself to tell my telomeres from age thirty-five, thirty-six, thirty-seven."

Even without that extra work, Angell said criminals can still defeat the DNA identification process—by creating a denial-of-service attack. "Felons will vacuum up DNA from football crowds and collect cigarette-ends," he said. "Low paid hospital staff will be compromised to supply hospital detritus: samples of blood, skin, saliva, and other biological material. Aspiring criminals, while perpetrating a crime, will randomly scatter an arbitrary collection of DNA material all over the crime scene, and the whole system will be compromised."[60]

Here Angell argues there's an economic cost when criminals can overload law enforcement. "In England, say all the criminals gather in this certain pub," he said. "What you do is go in and collect all the cigarette butts, and you go out and rob a house. You drop cigarette butts everywhere [at the crime scene]. Every butt that the law enforcement picks up has got a criminal record. They've got to check up on every one. So it's all down to economic cost. If the cost of doing all this work is so enormous, they'll give up in the end." This is, in essence, the hardware hacker's point of view: Find a vulnerability in the way we view the world, and exploit it.

Knowing there are vulnerabilities with biometrics, Marpet would prefer to see a variation on two-factor authentication. "It's going to get to the point where we need three-factor authentication," he said, with biometrics as that third factor. "The point is," said Marpet, "that third factor will support one leg of a three-legged stool: something you know, something you have, and something you are. It won't be a finger that gets me in everywhere. I'll still need my PIN code, my RSA token [a device that generates one-time passwords for access] code, and my finger."

If a thief cuts off your finger, he might have your fingerprint or your blood to squirt at the machine, but if the criminal kills you, he won't have your PIN code. "If any leg of that triangle goes away, you're done," said Marpet. "I'm fully in favor of biometrics as part of an integrated or layered security solution. I'm not in favor of biometrics as anything less."

Marpet added that the psychology of biometrics is really interesting "because people really believe this stuff. They ignore the error rate; they ignore everything. They totally ignore any objective data for the subjectivity of let's let a machine do it. The systems are expensive and hard to calibrate. They're not a magic bullet." Marpet laughs. "We really are trying to turn everything over to SkyNet [from the *Terminator* movies] as fast as we possibly can."

The fundamental flaw with biometrics is that people are not perfectible, and neither are the systems we design and call biometrics. "Technology imposes structure on our actions, which gives us at best a tenuous handle on uncertainty," said Angell. But it does not resolve uncertainty.

So, if current gadgets and technology do a poor job of recognizing us, might they do a better job of telling us something else about ourselves?

CHAPTER SEVEN

Zeroes and Ones

Think about your day. Not just the various appointments scheduled in your day planner or even the common tasks that we all perform, such as getting up, eating breakfast, going to work, lunch, and the gym, and so on. Try to think about the specific details of your day, such as the number of people waiting in line with you at Starbucks or the make and model of every car you passed on your way to and from your job. It's hard to recall such details. It may be impossible for most of us.

The human mind has evolved to record only the most significant details, such as the presence of a predator or specific places of danger to avoid. We simply discard the rest as noise. This makes sense; we can only focus on so much information at any given time. But what might we be losing in the long term? For all the evils of data collection explored in Chapter 4, might there also be some good? For example, do we all perform certain unconscious habits that might help us understand and improve our social interactions?

As we have already seen in previous chapters, personal data can and does leak out in unexpected ways. Stored or intercepted credit card and Social Security numbers become faceless zeroes and ones bought and sold on the Internet; radio frequency identification (RFID)–enabled driver's licenses and passports wirelessly provide tech-savvy criminals with biomedical and personal data to create new identities. An emerging class of personal data, records of where we go and even whom we associate with, however, is much less quantifiable. This too is noise, in part because it exists only deep within the vast collections of electronic data generated by our cars, worker-access badges, and always-on mobile phones. What if, using only a collection of the various forms of electronic data generated throughout your day, someone could predict your day-to-day and person-to-person interactions with 95 percent accuracy? And what if those predictions could further be combined with those about others to forecast the economic health of an organization—say, whether or not a company will be profitable or a community sustainable?

For the past ten years, Massachusetts Institute of Technology (MIT) professor Alex "Sandy" Pentland has been collecting and mining secondary electronic data, seeking to reveal otherwise hidden information about us as individuals and as society as a whole. "We are in the midst of an explosion of information about people and their behavior, but most of it is noise," admits Pentland.[1] His concept, known as reality mining, attempts to parse what's truly meaningful and find what he calls honest signals. The concept of honest signals, which are unconscious patterns of communication in humans that predate language, was named one of the Harvard Business School's Breakthrough Ideas for 2009.[2] His research has so far quantified some of these ancient patterns that sociologists have known about for years. And his hard data may help explain, for example, why some people are more successful at negotiation than others.[3]

In one test, Pentland and another researcher observed practice salary negotiations between mid-level executives transferring to a new company and their bosses. When strong mimicry was observed—that is, when both parties mirrored each other's minor head and body movements and speech patterns—the negotiations ended with both parties feeling satisfied. Another study focused on consistency; if, say, a boss announced during the salary negotiation something completely unexpected, the other person's speaking pace, emphasis, and even hand and body movements all became uneven as the brain worked on this new problem while at the same time tried to maintain the conversation.

Behavioral sociologists will tell you that neither observation is new; both mimicry and consistency are well-known behavioral concepts. So is the fact that a depressed person speaks more slowly and an agitated person speaks more rapidly and is more likely to cut another party off mid-sentence. Speech data can also shed light on a person's state of mind: A focused individual speaks clearly and deliberately, while a confused individual mumbles and hesitates.

Pentland has used electronic gadgets to automate his analysis. For example, mobile phone microphones can be employed to determine a speaker's emotional tone by using software to analyze how often a person interrupts other people and how long that person talks compared to others. Other gadgets can be used to record interactions during business negotiations.

To test his predictive theories, Pentland and his team often enlist up to one hundred subjects from the MIT Business School and Media Lab. In one nine-month study in 2004 and 2005, subjects were given use of various gadgets, including a sponsor-provided Nokia Series 60 mobile phone, preinstalled with several pieces of custom software. To gather location information, the MIT team mostly relied upon Bluetooth, a localized wireless signal enabled by default on most mobile phones today. Bluetooth is often set to "device discovery mode," meaning the phones broadcast their unique ID data across short distances in search of other Bluetooth-enabled gadgets. Even without Pentland's custom software, someone could use Bluetooth monitors in an office building to track employees with Bluetooth-enabled mobile phones—perhaps to track how many trips to the break room the employee took each day.

Pentland tracked his student subjects' movements every second of every day around campus in order to figure out what an observer might glean from all that data. In the end, his team recorded more than 350,000 hours' (or roughly the equivalent of forty years') worth of data just for one study.[4] Not surprisingly, the project provided a remarkably accurate and intimate view of each subject's life. Analyzing the data, the research team at MIT was able to look at how predictable people's lives were from day to day (what locations they frequented, whom they talked with and for how long). In particular, the team looked at how

organic social networks evolve (i.e., how people first meet and then continually meet up again and again). The researchers said they identified
the "characteristic behavioral signatures of relationships" and could
"accurately predict 95 percent of the reciprocated friendships in the
study."[5]

Indeed, an analysis of the data collected in the 2004–2005 MIT study
reveals not only time-stamped locations but also whenever names appeared within the phones' contact lists and the identities of those individuals being called. This further allowed the researchers to reconstruct
elaborate webs of personal interactions that had occurred throughout
the year and to identify the various subgroups operating within. For
that, the MIT team created a rudimentary program that predicted
friendships by "picking up the commonsense phenomenon that office
acquaintances are frequently seen in the workplace but rarely outside
the workplace."[6] For example, being near someone by the coffee machines in the break room at 3 p.m. conveys a different meaning from
proximity to that same person at a local bar at 11 p.m.: Friends are often
seen outside the workplace, even if they are also coworkers.[7]

Pentland took the proximity patterns during one week and exam-
ined the behaviors of self-reported reciprocal friends (where both persons reported the other as a friend), nonreciprocal friends (where only
one reported the other as a friend), and reciprocal nonfriends (where
neither indicated the other as a friend). A "social behavior pattern" is
the daily pattern of movement between the places where a person lives,
eats, works, and spends off-hours. Pentland's reality-mining research
shows that a majority of an individual's activity often falls within a
small repertoire of these social behavior patterns. Pentland's team
found that the behavior of each subgroup—for example, the amount of
time spent in proximity to others—differed, whether or not there was
communication between parties. Upon further analysis, the differences
could indicate each participant's social network of friends and coworkers with 96 percent accuracy.[8]

So far, we've only seen application of this within academic circles.
Do any of Pentland's ideas have application in the real world? Imagine
suddenly learning the name of, or having a conversation with, a stranger

you've seen on the train every morning for the last few years. What if the morning commute is just one of several points of contact you have with this individual during a normal day (you visit the same Starbucks at different times, you frequent the same grocery store on different days). Perhaps you even know one or more people in common.

Certain social-networking applications today report the location of their members. To preserve a user's privacy and minimize disruption, however, Pentland proposes a feature that would send an anonymous text-message alert saying only that a person with similar interests was nearby. Both users would then have to respond "yes" before they could share any personal information.

Foursquare is doing something like this already. Launched in 2009, Foursquare links registered users and updates their locations on Twitter and Facebook. Subscribers earn points for checking into venues in over one hundred cities worldwide. The number of times a person checks in, with only one visit per location per day counted, matters: Subscribers who check in more than anyone else within the last sixty days are named that venue's "mayor." The social element is that Foursquare notifies subscribers of friends who are nearby. There are commercial aspects to this. For example, Zagat, the restaurant guide, has signed a deal with Foursquare to create a "Foodie" badge to be earned when checking into Zagat-rated restaurants in major cities.[9] In response to the popularity of Foursquare (which accrued half a million subscribers in its first year), Facebook added a location-based service called Places.[10] However, there's a matter of scale. Facebook, with its 500 million subscribers worldwide, has already attracted thieves who use Places to see when people aren't home.[11]

If we use technology to start meeting strangers encountered throughout our day, then in some ways we'll be returning to our ancestral roots. Throughout history, human beings lived in tribes and villages, tight-knit communities where we knew every member within our community and they knew us. Urbanization has fractured our familial and even cultural roots. Yet, the concept of technology uniting rather than dividing should not be shockingly new; even in the early days of the telephone, some neighborhoods had party lines, lines shared by multiple parties; one could lift the receiver and listen in on someone else's phone conversation.

Today, online social networks inform any one viewer of everyone else's personal business. By having our mobile phones identify those around us through systems that connect us to others with similar interests anywhere in the world, we are becoming an electronic village, and, like denizens of those early villages, we may conclude, as did Thomas Malone, professor of management at the MIT Sloan School of Management, that "privacy is just an anomaly."[12] In other words, for most of human history, we did not have much personal privacy.

As we will see in this final chapter, electronic data can sometimes be better at telling us what is happening in the world around us than our own senses, and in some cases our privacy trade-offs may be in our best interests.

1.

To better understand the future, we need to understand the past. Pentland, in his speech at Yahoo! Labs, took issue with the Scottish Enlightenment, a period in the early eighteenth century when rational explanation for human behavior began to take root in Europe. Sir Walter Scott, David Hume, and others at the time said that man could affect his destiny if he better understood his behaviors. That's not always the case. "That's BS," says Pentland. "Not sort of right. It's wrong."[13]

Pentland talks about the brain as having two parts. There's the conscious brain that we're most aware of and that drives most of our action. The other brain, or the "automatic mind," is often talked about in a denigrating way. Pentland feels this is unjustified and asserts that this other part of the brain is at least as powerful as the conscious part. If you don't think about something directly, sometimes you make better choices. If you sleep on a problem, sometimes you make dramatically better decisions the next day.

The role of the subconscious brain becomes clearer when you look at prelanguage human societies. Human language is about 200,000 years old.[14] Yet, societies existed before that. What did early humans do before oral language? They had tribes, they were social, they hunted, they went to war, they settled, and they reproduced. Early humans did all that without words, without books, without information storage. They did it using their automatic minds.

Consider that some apes plan, invent tools, and do many of the same things that early humans could. And it's not just mammals. There are other social, nonverbal species—for example, bees. Bees use dance to communicate. Hives make major decisions that way. Worker bees go out to find food, and if one comes back and wiggles, other bees follow the first to find the nectar. If more come back and wiggle, then more bees go out. If the food source is really good, the bees may collectively decide to move the entire hive. In this case, amplitude and frequency do the communicating.

When humans invented language, Pentland says, language didn't replace earlier communication methods but instead built on top of existing routines. Pentland uses a term from evolutionary biology, *honest signals*, to explain what's operating below the surface of perception. He says these signals could explain why some people coordinate groups better, earn more money, and appear to be more successful. Honest signals are natural and hard to fake. They evolve in every social species, and only now do we have the digital gadgets in our midst to collect human data for further analysis.

The idea has been out there for a while. In his second book, *Blink*, Malcolm Gladwell writes about how only thirty seconds of observation can lead to accurate predictions of human behavior. Specifically, he cites the research of John Gottman from the University of Washington. By using what's called thin slicing, Gottman can observe a couple in a room, talking about any random subject, and within a few moments determine whether they will stay together or separate.

Gladwell says that certain behaviors within all of us, if you know how to look for them, can reveal much about a person, a couple, and even an organization. Gottman in particular looks for something he calls specific affect, or SPAFF, and ranks up to twenty different categories with numbers. It's tedious: "The notation '7,7,14,10,11,11,' for instance," Gladwell writes, "means that in one six-second stretch, one member of the couple was briefly angry, then neutral, had a moment of defensiveness, and then began whining."[15] But the couple wasn't just sitting in a room; they were also wired to various electrodes and sensors. Using all of that equipment, plus observation, Gottman looks for activity, influence, memory, and consistency.

Such minute observations, collected tirelessly through automation, may be beneficial to health-care providers. For example, in the realm of mental health, mobile gadgets can help therapists design better treatments for their patients. Accelerometers, which determine the orientation of the mobile gadget, might reveal fidgeting, abrupt or frenetic motions, and even pacing. Location tracking may reveal unhealthy obsessive patterns or extreme routinization. Decreases in movement, velocity, and range have also been linked with depression. Depressed people sound flat, exhibit low activity, have low empathy, and are inconsistent.

By collecting electronic data, some of which are the electronic bread crumbs we're already generating today, Pentland thinks he can find the honest signals in all of us. And he may be right. He has dug into the noise we produce every day to find the unconscious signals between two or more people.

With this data, Pentland can examine companies, cities, and societies. Without listening to words—by focusing on actions, no matter how minute—Pentland can predict the success or failure of each group. Using gadgets already available to us, he produces quantifiable data in the areas of activity, influence, mimicry, and consistency.

Activity, according to Pentland, is the amount of energy we expend in a conversation, which can predict depression. It is autonomic and involuntary, occurring just below the level of conscious thought. Increased energy associated with activity relates to the amount of interest we have in the social interaction; obviously, decreased energy relates to our lack of interest. Think of a child's energy level leading up to a birthday or Christmas. Adults can control themselves better. We adults think about the words we use in a conversation, the facts we want to share, and we monitor the other person's reaction to what we say before proceeding. All this takes energy, and the more engaged we are, the more interested we are in the topic being discussed.

Influence, on the other hand, is an activity that gets other people to match your pattern of speaking. It is thalamic, referring to the part of the brain situated between the mid- and forebrain that relays sensations and motor signals. When people pay attention to you, they respond quickly. If someone doesn't respond when you talk with them, you feel bad—you know they aren't paying attention. Thus, people who are considered dominant in society control conversations; people who are less dominant do not. During normal conversation, subtle differences in speaking occur within milliseconds. They are virtually impossible for the conscious mind to monitor and require the brain's attention and orienting systems. Here, ancient systems kick in reflexively. Pentland says that influence can apply to speaking pitch; the political candidate who controls the overall tone of debate is seen as dominant and therefore goes on to win the election. Audio from a mobile phone can monitor this.

Mimicry is the unconscious act of mirroring another person. Mirror neurons cause us to react when we see someone else reacting. For example, in a job interview, when a person leans back in a chair or tilts her head in response and the other person does the same, that's mimicry. Humans and apes posses mirror neurons—with them we can start to mirror another person (nod, cross arms). Even a three-hour-old infant can stick out her tongue if the mother does that first. Mimicry correlates strongly to feelings of connection and trust. A psychology study from Stanford used computer-generated avatars to explain the university's student ID card program.[16] Some students copied the actions of the animated character on the screen (tilting the head, leaning forward), not realizing they were doing so. When later asked about the overall message of the video, these students tested better than their peers who did not mimic the cartoon character. Pentland says mimicry is the unconscious signal for empathy. "If there's more mimicry," said Pentland, "there are higher levels of trust."[17] Motion sensors within a mobile phone handset can monitor for mimicry.

Consistency, then, refers to the volume and emphasis we give to our words. Expertise conveys confidence. As we gain experience in something, our words become more meaningful. For example, if we're in a negotiation, says Pentland, and the other speaker presents a new piece of information, our brain has to pause to deal with it, and this can slow the rate of our speaking. When several conflicting commands come down from the brain, it is hard to act in a consistent manner. Both audio and motion sensors within a mobile phone can monitor for consistency.

Already customer-service or call centers have taken advantage of this information. Within the first ninety seconds of a customer call, some call centers can determine, using special software, whether the call will be successful. The software does not listen to the actual words spoken but to energy, tone of voice, and other qualities of the two speakers: "Oh, really? Tell me more about that." It also listens for mimicry: "Uh huh." Those call centers that analyze and act on customer interactions have shown higher customer satisfaction, according to Pentland.[18]

Although it is doubtful that anyone today is harvesting and reality mining all the random data we produce, Pentland believes firmly that we each should have the right to opt in to such collection. But as we have seen, auto manufacturers, transit authorities, and even the National Security Agency have started collecting random data about each of us without our consent. And as storage gets more inexpensive, there's reason to think that other private companies will start collecting more of this electronic background noise as well. Some organizations already have, albeit for the public good.

2.

Driving the freeways near Washington, DC, sometimes requires a leap of faith. As the belt of highways encircles the nation's capital, you cannot always see what traffic lies beyond the next bend. As humans, we're limited by the natural senses in our bodies; we simply cannot see the larger picture without collecting electronic data. To aid the driver, transportation agencies in Maryland and Virginia have been using technology that reads the unique identifier transmitted by all mobile phones to map and report up-to-the-minute traffic flow on digital signboards and other media.[19] The phones need only be turned on, not in use. By using mobile phones to provide a constantly updated picture of traffic flow across thousands of miles of highways, transportation agencies can spot congestion and divert drivers by issuing alerts via radio or electronic road signs.[20]

Of course, transportation agencies have long had the ability to monitor traffic flow without the use of mobile phones. But the data from these early traffic-monitoring systems was spotty, subject to omissions and gaps and often limited to certain troublesome stretches of road. In short, traffic engineers were unable to capture what was really happening across the entire highway network. That's because the monitored stretches first had to be installed with individual sensors embedded in the concrete—entailing not only tremendous installation costs but a continual maintenance cost over time as the equipment wears out or needs to be updated. Thus, only critical stretches of road could be monitored. The sensors are additionally supported by a series of remote television cameras that also need to be maintained.

Mobile phone users move around constantly, and so the gadgets broadcast a silent beacon looking for the best reception as they shift from cell tower to cell tower. This has its pros and cons. This back-and-forth negotiation between the mobile phone and its nearest towers produces the number of bars, or signal strength, displayed on the handset. It also does something else: It leaves a footprint. Cell towers record the time at which a person's mobile phone first comes into range and then exits—even if the phone isn't being used to make a call. For this process mobile carriers use only the phone's unique ID numbers to contact the cell towers; personal account information about the user (name, billing address, method of payment) is kept back at the carrier's offices and never broadcast.

At some point, someone realized that a cell tower's unique ID entry and exit data can be triangulated with other towers, plotted in aggregate, and displayed against highway maps. In other words, this system can track the mobile phones (whether they are being used or not) in people's cars as the cars move down a highway, passing from one cell tower to another. The time it takes the phone to enter and exit each cell tower's range can then be correlated to the road speed.

The resulting map shows the traffic speed along major and minor arteries at considerably less cost than traditional traffic-monitoring methods.

For the convenience of a faster commute, this system currently poses little risk of privacy violation. Transit agencies typically do not store the tremendous volume of fleeting data. Besides, not every road is serviced. The mobile phone–based traffic systems in use in California, Georgia, Maryland, and other states are regional. A project in Missouri, however, is statewide, monitoring every road, even in rural areas with significantly lower traffic counts. It is the Missouri system that has given some people pause.

At its inception, privacy rights advocates campaigned against Missouri's statewide system, arguing that mobile phones' unique ID do not provide enough of a firewall to protect drivers' privacy. "Even though it's anonymous, it's still ominous," Daniel Solove, a privacy-law professor at St. Louis–based George Washington University, told the Associated Press. "It troubles me because it does show this movement toward using a technology to track people."[21]

In Missouri, privacy rights advocates frequently cite the following hypothetical scenario: Suppose a state trooper noted one cellular phone user's unique ID traveling at excessive speed along a highway; that alone would be probable cause for the trooper to obtain a court order for the carrier to release the cell phone owner's personal account information. The state trooper could then issue the owner of the cellular account a speeding ticket.

The privacy-violation outrage here is probably unwarranted. The effort required to track down the account information for a single cellular ID speeding in the middle of a rural highway isn't worth the reward of the nominal fine the ticket produces. Although the ability to place all this cellular ID data in a searchable database certainly exists today, unless a city or state can find a way to achieve a high return on investment for its efforts, it's unlikely that we'll see virtual tickets any time soon.

But perhaps a more intriguing question is whether the international mobile subscriber identifier (IMSI), the unique ID, can be a source of information for a third party about the owner. Despite what mobile carriers say, the IMSI can be used to glean sensitive information, such as your identity. In Chapter 3 we talked about IMSI catchers used by some law enforcement agencies. Here we're talking about deconstructing the information within the ID to identify the owner.

Let's look at another seemingly trivial piece of mobile phone information that has been used to gain personal information. To assist Apple iPad early adopters, AT&T created a website that required them to enter a number, a nineteen-digit integrated circuit card ID (ICCID). This number, uniquely associated with the iPad's subscriber identity module (SIM) card, enables cellular and therefore Internet access. After subscribers entered the number, the screen would populate with the e-mail address they provided when they activated their iPads.

Security researchers at Goatse Security suspected there might be a flaw with this system, and so they produced an automated script that input random ICCIDs to the AT&T site to harvest the e-mail addresses of individuals who had purchased and therefore activated their iPads— capturing data for 114,000 iPad early adopters.[22] The group did not go public with the flaw until AT&T changed the website. Included in the published list were e-mail addresses for White House Chief of Staff Rahm Emanuel, New York mayor Michael Bloomberg, and *New York Times* CEO Janet Robinson.

AT&T assured the public that integrated circuit card IDs (ICCIDs) could not reveal much more about users; indeed, this flaw required the use of another database to reveal the e-mail addresses. However, several research projects came to light showing that the nineteen-digit ICCID could be used to derive the more important IMSI, especially in the way that AT&T chose to assign its ICCID. Although the ICCID does not reveal much useful data, it does put an attacker one step closer to acquiring personal information.

The system carriers use to manage their mobile networks is Signaling System 7 (SS7). It includes a protocol suite, as well as switching and tracking hardware and software. At the Chaos Computer Conference in 2008, researcher Tobias Engle deconstructed the SS7 protocol in order to discover where a mobile user was at any given moment.[23] Critical to this process was the string of data identifying the mobile switching center

(MSC), which handles voice calls and Short Message Service (SMS) text messages. Actual location is harder to determine as larger cities may have several MSCs, but the location of the physical cell tower is usually a good indication of what city and country your SIM card is in.

Engle first automated the process of finding existing cellular numbers. He had a computer system randomly call people at night. The call didn't have to go through; just one ring or two was enough to send back the current MSC information for the number. He and his team had the idea that if they plotted enough MSCs for people whose locations they knew, they could build a database (a cellular white pages) to find the people whose locations they did not know. By mapping these locations in aggregate, the team found common properties among the MSCs, such as type of service and numbers assigned. Any outliers were mobile users traveling outside their home area.

In theory, only network operators should know where you are (this, in order to route calls to your phone correctly). Unfortunately, noncarriers or nontelcos today offer lookup services similar to Engle's. They make a call to retrieve the MSC and IMSI information. As one consequence, suggests Engle, an evil spammer could enlist these companies to obtain MSC information, then direct spam to that network's SMS center via SS7, resulting in hundreds of spam SMSs sent for free due to a quirk in network design.[24]

One solution is called home routing. Currently, outgoing SMS text messages pass through what's called the home network before going out to the recipient. But incoming SMS text messages do not pass through the home network; they come directly to the handset, which is one reason why you receive SMS spam. The process of home routing sends both outgoing and incoming messages through the home network, allowing the carrier to filter incoming spam.

Such a home routing network would also allow the recipient to keep an archive of all SMS text messages sent and received (since they would both pass through the home network). Home routing has been adopted by mobile carriers around the world, some of which are currently offering mass spam filtering at the hub. It has yet to take root in the United States.

Can location still be determined without access to SS7 information?

As noted with traffic-speed monitoring above, the location of any caller can be determined with as few as two cellular towers through what's called triangulation. One method, called angle of arrival, looks for points at which the lines of communication with two different cell phone towers intersect. In urban areas, where cellular towers are more numerous, this method is very precise. A variation, called time difference of arrival (TDOA), requires more towers. Here, the location can be triangulated based on the time differences between signals reaching the multiple towers. AT&T and T-Mobile currently use TDOA. The problem with both of these methods is that line of sight with the cell tower is necessary, which is not always possible, for instance, in tunnels, underpasses, and mountainous or rural areas.

The Carmen Sandiego Project has looked at what information is available on a GSM network, such as the Mobile Subscriber (or Station) Integrated Services Digital Network Number, which is the mobile phone number assigned to the SIM. The researchers produced a wealth of data on the value of each of the signals our mobile phones produce every day.[25] Researchers Don Bailey and Nick DePetrillo have found ways to identify users just from basic information like an IMSI. Speaking at SOURCEBoston in 2010, Bailey and DePetrillo attempted to re-create Engle's work without access to SS7. For example, using commercial lookup services, which send and receive current data about the cellular owner, they found the MSCs (which tells them the closest cell tower, which gives them a rough idea where a subscriber lives) for people they knew were home—by requesting the information at 2 a.m. Like Engle, by knowing the MSC, they could then try to find the locations of people they didn't know—for example, Tiger Woods.

Bailey and DePetrillo also used the caller name, or CNAM, listings to find corporate listings—for example, the number for Apple Computer Inc. in Cupertino, California.[26] By looking at the North American Numbering Plan Area (NPA) registry and local number prefix (NXX), they learned some details about the number, such as the provider of the cellular service (at least the provider when the number was first issued). One commercial site I tried did not correctly match the zip code with the cellular number I tried, but other information (latitude and longitude of the city) was correct. Another commercial service, in Europe, can perform up to 100,000 MSC lookups for a mere €110.

Knowing the NPA NXX, therefore the mobile phone number, gets you closer to knowing the current location of that phone. Bailey and DePetrillo could now look for instances of specific phone numbers on the Home Location Register (HLR). The HLR tells the mobile network where a subscriber is at any given moment. This is important for incoming calls. Since the mobile device is . . . well, mobile, the network must consult this database to know which base station to route the incoming signal to. When the subscriber goes off the network—for example, leaves the country—an additional database is used. The Visitor Location Register (VLR) is used for any kind of roaming outside of the home network, international or domestic. In practice, that usually means international, but not necessarily.

Without carrier resources, Bailey and DePetrillo managed to take additional bits and pieces of mobile information (such as the IMSI, CNAM, and HLR) and put together a mobile white pages that would allow them to stalk someone.[27]

Further, any phone signal seen moving from base station to base station might indicate that an individual was driving. That information could be linked to traffic cameras. Bailey, who lives in Denver, noted that Colorado makes live traffic images accessible to anyone via the Internet. Now an attacker, perhaps using Bailey and DePetrillo's method for knowing a person's cellular phone number, might also learn the make and model of a victim's car, perhaps even its license plate number. Using additional databases, the attacker could follow the license plate to a home address and obtain other personal information.

3.

Cell phone records may play a positive role in public health. We have seen that cellular records can be kept indefinitely and how that might be a privacy problem. On the other hand, the existence of such records, if the personal data is removed, may contribute to the common good.

For the last few years, the local government of Zanzibar has struggled to diminish the number of malaria infections among its 1.2 million residents. Malaria is caused by a parasite transmitted in the saliva of certain species of mosquito. Humans can't transmit malaria to one another (except through blood transfusion, perhaps), but if a mosquito bites a person infected with malaria, the insect can then spread the disease to other people it bites.

The antimalaria program currently in place in Zanzibar uses pesticides and mosquito nets to kill mosquitoes or keep them from biting humans. Malaria-carrying mosquitoes tend to be more active at night, hence the use of netting over beds. Infection rates in Zanzibar, once as high as 40 percent, have dropped to less than 1 percent today. The country is banking on the cost of malaria eradication being less than that of managing the control measures currently in place. Threatening that eradication, however, is the fact that many Zanzibaris travel off their two islands and head to mainland Tanzania, where the risk of acquiring malaria is much greater.[28] The typical mode of transportation is a four- to six-hour ferry trip between the mainland and the islands. But ship manifests, even basic records such as when the boat left the dock and when it returned, are not well maintained, and this is a problem because health officials have no way of monitoring how many people visit the most malaria-infected regions of Tanzania.

So students at the University of Florida came up with a unique monitoring program that uses cellular records to help track malaria transmission in Zanzibar. The students have been studying data provided by the Zanzibar telephone company to arrive at accurate predictions around the spread of the disease. They were given the records of nearly three-quarters of a million customers whose movements could be tracked using a computer.

Most cellular customers never leave Zanzibar. The students could see this because their numbers were not linked to calls made from within mainland Tanzania, and so they were quickly eliminated from the study. Some phones, however, made frequent calls from Zanzibar on the first three days of the week, then from mainland Tanzania later the same week. Most of these users, the students found, traveled to the safe city of

Dar es Salaam, but a few hundred traveled outside that city into part of Tanzania where populations have an infection rate of up to 40 percent.

Like other studies, the University of Florida study concluded that the government could give residents preventative medicine before they traveled to the mainland or screen all residents for malaria upon their return from Tanzania. Both are expensive options.

Alternatively, the University of Florida recommended the country of Tanzania could target its malaria-prevention information campaign to the few hundred high-risk travelers who ventured past Dar es Salaam, thus saving the government money and achieving its goal of eradicating malaria within its borders.

Pentland also sees health-care opportunities in mining mobile phone data.[29] For example, most medical studies track the overall population. With mobile phones, tracking subpopulations might be better for disease control. As we saw in Zanzibar, not everyone who rides the boat to mainland Tanzania needs to be targeted with antimalaria information, only the subpopulation that travels beyond Dar es Salaam. Already Pentland's research within the United States suggests that most people have only a small repertoire of behavior patterns (where they live, eat, work, and relax), which account for the majority of each individual's activity. Thus, one can cluster similar behavior patterns and examine how infectious diseases are communicated or when individuals might face the greatest risk of exposure. Pentland has found that, in general, subpopulations tend not to mix, so understanding the unique characteristics of an at-risk subpopulation should help public health services improve public health education and intervention.[30]

Pentland thinks that his reality mining could further be used to determine not only whether the health intervention was working but whether the underlying theory behind the intervention was correct. Pentland says it would also be good to know if the intervention simply failed or wasn't adequate in the first place.[31]

Reality mining can also be used to better understand why some chronic health-related conditions appear to be contagious within subpopulations. Actually, these conditions generally stem from habits and behaviors. Two obvious examples are lung cancer (smoking) and obesity (overeating, poor nutrition, lack of exercise). One theory holds that an individual's health is linked to that of others within his or her immediate social network, which makes sense considering that we tend to mimic the health habits of our friends and family. If this is true, Pentland suggests targeting individuals in at-risk populations.[32] Such theories should be tested, however, with an informed and consenting test population, such as Pentland's students back at MIT.

4.

In the United States, the number of people aged sixty and older will number 108 million by 2025.[33] The existing retirement communities cannot handle such an influx. Fortunately, many of these adults will choose to remain independent. And it's Dr. Jeremy Nobel's opinion that technology, namely gadgets, will play a vital role in their continued independence by remotely monitoring their health and safety.[34] He's not alone. Eric Dishman, chief strategist and global director of product research, innovation, and policy for Intel's Digital Health Group, agrees.[35]

The switch from the current system of providing health care in hospitals and senior facilities to delivering more powerful, personal health care in the home would be huge. Dishman talks about meeting the "age wave" head-on with technology produced within the next ten years.[36] He cites a gadget called SHIMMER, which stands for Sensing Health with Intelligence, Modularity, Mobility, and Experimental Reusability. It fits into a pocket and can monitor hand tremors, gait, and stride length. Another gadget, called a Magic Carpet, embeds sensors in a carpet to track a person's movements. A break in daily routine could signal that new medication may not be working or that the person has become injured and immobile.[37] Having a health-care framework encourages innovation and easy adoption.

Back in 2002, a team of Dutch researchers partnered with Ericsson Consulting in Germany to outline what they thought a health-care network would look like in the coming years. They focused on wireless technology, USB gadgets, Bluetooth, and the use of 2.5G and 3G mobile services. They envisioned what they called a wireless body area network

(BAN), where patients at home would send vital statistics to their health-care providers, along with video and audio. This study was done long before the creation of iPhone apps or today's faster mobile networks. The researchers did, however, envision the need to build security and encryption into the system rather than bolting it on after the product's release. Security is not always discussed in current e-health debates. They also recommended that various BAN technologies be integrated into mobile gadgets, including watches, which would be worn by the patient every day.[38]

In addition to Bluetooth technology, which we discussed in Chapter 3, the Dutch-German team considered the short-range radio technology work being done by the Zigbee Alliance.[39] This is a group of communication companies with systems that need to be securely networked but also produce low data rates and require long-term battery life, such as remotely operated thermostats, energy monitors, and smoke and fire detectors. These same requirements would have application to health care.

The Dutch and German researchers were more conservative than Nobel and Dishman, however. They imagined only scenarios in which chronically ill patients could be remotely monitored, minimizing the need for continued hospitalization. "Early discharge patients" were those requiring intense monitoring to prevent readmission to the hospital. A second monitoring category included those patients with respiratory illnesses exacerbated by another life-threatening illness or comorbidity.

Underlying more ambitious gadgets such as SHIMMER and Magic Carpet are new technology platforms such as the BioMOBIUS ecosystem from Technology Research for Independent Living.[40] This is an open technology platform providing for, among other things, mobile hardware and physiological sensor support, biosignal processing applications, and graphical standards, including graphical user interfaces.[41]

Dishman envisions these new technologies will help shift 50 percent of all current hospital care to the home. He describes remote health-care monitoring as a disruptive technology, improving services for the market in ways we cannot yet predict.[42] As for who will bear the costs or what industry standards group will win out, he doesn't say. "No matter who pays for it, we better start doing [health] care in a fundamentally different way," he said.[43]

Remote monitoring of patients is known as telehealth or telemedicine. As we saw in Chapter 2, the ability to make medical equipment, such as a pacemaker, accessible via the Internet opens gadgets to remote attacks. Nobel and Dishman are talking about a more secure future by building on secure frameworks such as the MobiHealth Network.[44] Another initiative is the Continua Health Alliance, which seeks to create and promote remote medical gadgets with its open standards.[45] The organization's design standards include Bluetooth and Near Field Communication for wireless and USB for hardware connectivity.[46] Participating companies include GE Healthcare, Medtronic, IBM, Cisco, Intel, Aetna, Nokia, Panasonic, Sprint, and Texas Instruments, among others. The products would be freed of some of the security concerns identified elsewhere in this book by requiring authentication for any changes to the firmware within the chip and encrypting any data entering or leaving the gadget. Such products would become "Continua certified." They would also have to be "future proofed" so as not to become obsolete too soon. And they must be easy to use. The group's first certified product, the Nonin 2500 PalmSAT handheld pulse oximeter with USB, monitors vital signs remotely.[47] By the end of 2009, six products had been certified, including remote blood pressure gadgets and a remote blood glucose monitor.[48] Some of these gadgets can record medical information and send it securely to a health-care provider.

According to Nobel, who coauthored a report titled "State of Technology in Aging Services: Summary," published by the Center for Aging Services Technologies,[49] a number of different health-related categories still need to be defined. One category is health and well-being, at the center of which would be electronic health records. The records would integrate the information from the remote monitoring with more traditional visits to the doctor's office. Older consumers, says Nobel, would decide whether to share this information with all of their caregivers. The technology would be designed to monitor older consumers and even to incorporate proactive steps such as simple exercises to maintain their current activity levels. Additionally, it would allow caregivers to observe the efficacy of treatments.

Safety is another category. Maintaining the ability to live independently has two facets: Not only do older people want to live on their own, but they should feel safe in doing so, knowing that if they fall, or otherwise fail to carry out a daily activity, the omission will be noticed and someone dispatched to help them. Nobel also talks about the ability to turn off stoves left idle or monitor bath water temperatures to prevent scalding. For dementia patients, more drastic means, such as locking doors to prevent late-night wandering and even satellite tracking, have been suggested.

Yet another category is social connectivity. Elder-care facilities offer social interaction, game playing, and other activities. Independent living often does not. In the future, home-based care may include passive monitoring for cognitive decline or enhancement of memory by getting the older person to talk about him- or herself. A simple solution is to enhance mobile phones and video telephones with various ease-of-use features so that older users can feel connected to the outside world.

As idyllic as this may sound, Nobel admits there are some very real barriers. One is economic. Not every elderly person will be able to afford this new technology, and it seems unlikely the federal government will pay for these services via Medicare. More important is the lack of awareness of such products in the first place. Older consumers may remember early efforts with hooking up home computers to the Internet and decline even to try newer technologies. And then there's pride: Older people may not believe they even need to be monitored and resist the technology on those grounds. Nobel points out that older consumers may also avoid asking for their own electronic health records, believing instead that their primary care physicians have all the information (some lack other doctors' records).

Lastly, there is no consensus on the value of these emerging technologies. Without that, Nobel says, there will be little investment and adoption.[50]

5.

Pentland, in his various writings and lectures, sometimes compares the world to a single organism, with mobile phones as its eyes and ears, sensing the world as never before. That simile works in ways that are both promising and disturbing.

Researchers at George Mason University pushed that analogy and wondered whether it was possible to enlist the millions of mobile phones and laptops around the world in a surreptitious surveillance network, dubbed "BugNET."[51] The effect would be to chain computers and personal gadgets together to create what's known as a botnet, a private network of compromised gadgets, which could then listen in on conversations, or call up mobile camera images, from any part of the world.

Such an idea is not new. In 2006, U.S. Department of Justice officials approved monitoring a pair of Nokia phones owned by two members of the powerful Genovese family, a New York organized crime family, in a federal district court decision known as *United States v. Tomero*. The FBI successfully argued to the court that members of the organized crime family were wary of phone traces and being trailed, hence needed different, more effective surveillance techniques.[52]

The investigation focused on John Ardito, a high-ranking member of the Genovese family. Investigators found that Ardito met regularly at a New York City restaurant, but concerned about listening devices, he also met at three additional restaurants. The judge allowed those other restaurants to be bugged, but the Genovese family found the devices. As the investigation progressed, investigators learned that Ardito no longer met with other criminals in person but spoke to them by mobile phone. The investigators applied for an exemption to the wiretapping provision of U.S. law (18 U.S.C. 2518[11][a]) and used what is called a roving bug, an application placed on a mobile phone that allows law enforcement to activate the microphone on the device even when the phone is turned off. This allows the suspect's phone to intercept conversations within its range.

The judge allowed investigators to bug the phone belonging to Peter Peluso, an attorney and close associate of John Ardito. Throughout 2004, private conversations were recorded, and in 2005, after being approached with the information, Peluso agreed to work with the government. The bug was removed, and Peluso wore a conventional wire.

It was those recordings that helped bring down John Ardito. Ultimately, forty-five members of the family and their associates were indicted on charges for crimes that occurred over twenty-two years. However, the family and associates filed a motion to suppress the information collected via the roving bug.[53]

Apart from the idea that our mobile phones could be eavesdropping on our conversations even when turned off, some find it worrisome that the warrant was issued without specifying an exact location. The Fourth Amendment states that a warrant must "particularly describ[e] the place to be searched." The FBI's use of a remote bug in this case will have to be decided by a higher court.[54]

But such a capability could also be useful. If technology makes us a global village, a single organism, roving bugs are a literal extension of that idea. People could tap into the mobile BugNET and, through geolocation, find an Internet-connected gadget near where they wanted to listen in or see what was happening, for instance, in the case of a historical event (a political revolution) or after a disaster of some kind. Rescuers could find webcams and mobile phones in the general area to get an immediate sense of what was going on near the epicenter of the disaster. This would require a degree of trust in government, which, in some parts of the world, is a cultural impossibility. Still, were we to find some accommodation, the use of our mobile phones for humanitarian good would be a boon.

6.

Pentland and others suggest we're entering into a new era, one in which we will start to recognize and accept different degrees of personal privacy. Perhaps we're simply done with the one-size-fits-all privacy model. In the future we might cede to a government electronic data for civic health and safety reasons and forego privacy in exchange for reduced prices on goods and services, yet resist calls to share our data in other cases—that is, provided we're given the right to opt in and out of data collection. If we are, the trick will be assessing the risks involved with each choice.

Pentland points people to the outbreak of Severe Acute Respiratory Syndrome (SARS) back in November 2002 and suggests the epidemic might have been identified weeks earlier.[55] More lives might have been saved had reality-mining techniques been used. "If I could have looked at the cell phone records, it could have been stopped that morning rather than a couple of weeks later," Pentland told the *New York Times*. "I'm sorry, that trumps minute concerns about privacy."[56]

Pentland is referring to Amoy Gardens, a high-density apartment building in Hong Kong, where, at the end of March 2003, an outbreak of SARS occurred; for one week all the residents living on one floor simply did not report for work.[57] No one reported this at the time, and two weeks later, on April 15, 2003, after the epidemic had become well known, the World Health Organization reported a total of 321 cases of SARS within that one Amoy Gardens building alone. Had the Chinese government used reality mining, officials might have been alerted early on that something unusual was happening to the people living on one floor of that complex.

Behavioral scientists tell us people tend to be less adverse to risks to their privacy and anonymity if there is some immediate benefit. It's also very much a cultural thing. People in Asia, for example, tend to be much more accepting of surveillance than people in the United States and Europe, where companies have known for years that people are usually willing to relinquish a portion of their privacy (name, address, e-mail) if they are doing so in exchange for some discount or gift. But is the result worth a potential breach of personal privacy?

There are ways we can avoid risks to privacy and anonymity posed by traffic monitoring. One obvious way is to power on your mobile phone only when you intend to use it. Any messages sent to you when the phone is powered off will automatically queue in your voicemail. And if you only make calls or check messages from one location every day—say, your home or office—not only will you avoid detection by someone's proximity monitor, but the resulting location data will be useless.

But this practice is inconvenient. Why even have a mobile phone? It also does nothing about anonymity. With mobile phones in the United

States and other countries, anonymity is almost impossible. The wireless carriers, operating as utilities, insist they are extending to you credit for a service and a product; therefore, they require you to provide your name, address, and Social Security number for a credit check.

Some companies, however, such as Tracfone, Virgin Mobile, Boost Mobile, and even T-Mobile, sell phones with prepaid service plans at such stores as Target. As long as you remember to purchase additional airtime periodically, you can have prepaid mobile service for as little as $20 every ninety days. Much like prepaid credit cards, prepaid phones are intended as an option for those with poor credit history. But if you pay in cash, they also provide an anonymous mobile phone option. You can also change your phone number often. These phones do, however, offer limited features and have a reputation with law enforcement for being used by those who don't want to leave a trace—namely, drug dealers and terrorists.

Transponders, like mobile phones, can also be prepaid. Systems like FasTrak and E-ZPass have unadvertised, anonymous account options for motorists who do not wish to be tracked. An anonymous FasTrak device, which still has a unique ID number and is still logged, is linked to a cash-only account with no personal details. The driver need only go to a service center and settle the account each month in person, which, for most of us, is inconvenient. And so, our attitude toward these passive tracking technologies might end up being a bit like our attitude toward sweets—the short-term benefits of relaxation outweigh the long-term effects of obesity.

Pentland thinks we need to come up with a new way of "being private in a world where there are sensors everywhere."[58] This may not be possible because the surveillance technology is already out there, and it's undetectable. If you try to create privacy by passing more laws, there will always be people who will break those laws. Privacy as many of us grew up knowing it is gone forever, thanks to technology (think of pinhole video cameras and the spyware that turns on the camera and microphone on a cell phone).

Still, Pentland believes that the user should own the information generated by the gadgets we use every day, and I share this belief. Pentland further believes that we have the right to state how and when we want that data collected by opting in or out of data-collection systems. You should also have the right to destroy, remove, or change the data as you wish. He calls this a "New Deal" on privacy in which we knowingly make a social contract with our gadgets.[59]

He suggests that companies that collect personal data should create an account, enabling you to see your data at any time. As data owner, you would have full control over the use of your data and could remove it at any time. Finally, as data owner, you would have the right to dispose of or distribute your data. If, say, you had data held by one company and wanted to redeploy it elsewhere, you should be able to do so.

In the age of such digital awareness, Jonathan Zittrain, professor of Internet law at Harvard Law School, facetiously suggests that humans should create a "nofollow" tag like that used in HTML pages on the Internet.[60] Such tags prevent robots or spiders sent by search engines from going any further and are used on Web pages that a company or individual does not want indexed. With our personal digital cameras and mobile phone microphones recording everything we do, a real-world "nofollow" gadget, like a radio frequency identification (RFID) tag worn on your belt, could broadcast your disapproval of this particular method of surveillance.

Though there is no legal obligation to honor your request, Zittrain's point is clear. You're asking another human to respect your right to privacy. With the rapid pace of global gadgetization, that privacy that we in the United States and Europe see as a basic human right is increasingly being lost. In Asia, where surveillance is part of some cultures, this is not as much of a concern. Might we be moving toward the middle where some surveillance is accepted and others not?

One solution might be use-limitation laws. Europe already has several strict privacy policies regarding commercial data collection; the United States does not. Under such use-limitation laws, anonymous IDs for mobile phone customers could be sold to transit agencies and others for monitoring purposes, but personal account information could not be sold or otherwise used. Further, the carriers and transit agencies collecting the detailed information would only be able to retain that information for two years. That would limit the period during which you could contest a mobile phone bill or even a toll transponder charge, but most people would probably be willing to live with that trade-off if aggregate, anonymous, and short-term use of such data could be used to produce benefits for society.

"When photos were introduced," Pentland said, "the natives in New Guinea thought you were capturing their spirits when you took a photo. It wasn't that different here. People had to get used to it. [With modern technology] we have to guarantee privacy and personal control so you are as comfortable with [the technology] as you are with having photographs of yourself."

In other words, whether we are ready or not, as a consequence of the gosh-wow gadgets in our lives, our attitudes about privacy and anonymity are about to evolve.

CONCLUSION

High Hopes

Perhaps nowhere is the speed of technological change more visible than in South Africa. As of 2009, there were 4 billion mobile phones in a world of 6 billion people, and some of the phones are in remote South African villages, where people like Bekowe Skhakhane live. Skhakhane, a thirty-six-year-old woman, fetches water from a river a mile away, lights her home with candles in the evenings, cooks over an open flame, yet relies upon her Nokia second-generation (2G) mobile phone to talk to her husband on a job in Johannesburg, 250 miles away. This raises an obvious question: If a village has no running water and electricity, how are people like Skhakhane keeping their mobile phones powered up?

According to the *New York Times*, a few miles down the road from Skhakhane, there lives a woman who will never own a car herself, yet keeps a car battery in her garage. She allows Skhakhane and others to recharge their phones for a few cents each day. Once a week this woman takes her battery to a gas station twenty miles down the road to keep it charged. Mobile phones are that important to the village: Skhakhane, for example, does all her bill paying and banking on her 2G mobile phone as well.[1]

By skipping landline phones and therefore dial-up Internet, several remote villages in South Africa have lurched into the twenty-first century, wirelessly interconnected with the world via residents' mobile phones. And these same people, who previously lacked access to banking institutions and even ATMs—financial services most of us take for granted—are now banking and purchasing via the mobile phone, leap-frogging the branch banks and online banking experience altogether.

Wizzit is an example of a virtual South African bank.[2] Previously a division of the South African Bank of Athens and now independent, Wizzit does not have ATM or branch locations, but it does issue a MasterCard-branded debit card. It also uses an older Global System for Mobile (GSM) communication protocol, Unstructured Supplementary Services Data (USSD), for money transfers.[3] Like Short Message Service (SMS) text messages, USSD was designed for quick, short messages and was not intended to support payments or financial transactions. Outside the United States, Europe, and Asia, the mobile phone population uses mostly legacy 2G stick phones with almost no Internet access. USSD is available on almost all mobile phones worldwide. In developing countries, payments are made via USSD-based text messaging because it is also relatively inexpensive to use. For example, Wizzit charges less than a penny for each transaction, much less than traditional banks would charge. Physical deposits are made at local post offices. Many of the mobile banking accounts are very small, too small for traditional banks, so despite the potential for fraud and theft, profit-oriented cybercriminals are not as interested in attacking these accounts.

And it's not just within South Africa. Farther north, the Kenyans are racing to construct enough cellular towers to handle the increasing demand on their mobile networks. The M-PESA mobile payment service has been expanding in Kenya since its launch in 2007.[4] M-PESA, which stands for "mobile money" (*pesa* means "money" in Swahili), uses SMS text messages and a "wallet" installed on the mobile phone's subscriber identity module, or SIM, card. According to the World Bank, only 26 percent of the households in twenty developing countries keep their assets in formal financial institutions.[5] The other 74 percent often do not qualify; in some countries, you need to be employed to have a bank account. Of those who do qualify, some feel intimidated when entering a bank or ignored once inside, or they simply can't make the trip via bus to a major city with a bank. Others are small business owners who do not earn enough to cover the costs of banking. Getting more people to invest in a financial institution increases the amount of capital available in that country to reinvest in projects such as homes, hospitals, and schools. Having a mobile phone links the unbanked to virtual financial institutions.

In the Philippines, over 3.5 million mobile bankers use either GCASH or SMART Money as alternatives to traditional financial institutions.[6] And in Brazil, mobile banking is overtaking traditional online banking, according to one Banco do Brazil representative.[7] He added that of the 2,000 daily m-banking transactions, 23 percent come from prepaid phone users, meaning they have no formal service plan with a given mobile carrier.[8]

Like Paula Ceely trusting her GPS navigation system despite her commonsense recognition that the virtual data might not match physical reality, the Third World is placing high expectations on mobile technology despite known risks. Although SMS text messages are not secure—they are unencrypted, so account information may be visible to man-in-the-middle attackers—a mobile commerce ecosystem is nonetheless developing within these countries, with shops, gas stations, post offices, and supermarkets accepting mobile deposits and payments.[9] This success suggests that the future of financial institutions is not in a brick-and-mortar location but in the Cloud.

Such innovation has expanded to rural health care, with physicians using mobile phones to instruct village women on neonatal care, parents on illnesses affecting their children, and the elderly too frail to travel for help. Mobile phones are even used to conduct basic annual examinations with additional kits containing blood pressure cuffs and stethoscopes. While some patients may still make the miles-long trek to see a doctor, they will do so less often.[10]

Rural African, Philippine, and Brazilian villages skipped over use of the online PC and proceeded directly to the mobile platform, albeit not even the most modern mobile platform. Unable to afford the latest technology, they sat out the first part of the Internet revolution while the rest of us worried about the number of applications on our desktops and gadgets or about the ease and speed with which we could get online or stream a video. Now they are taking older technology that we upgraded past years ago and putting it to practical use—not streaming video but paying bills and seeking medical assistance when needed.

All of this innovation should fundamentally change how we view our mobile gadgets. Yet, before the United States and Europe can reach that level of trust in their mobile networks, a bit of work remains to be done.

There are many reasons why mobile banking has not taken off yet in the United States. In Africa, the Philippines, and Brazil, villages have one mobile network and generally one large, central bank, making such a transition easier. In the United States, the mobile telecommunications giants, the banks, and software providers haven't agreed how to work together and make money. Internet innovation only started when charges for Internet service provider data usage became flat monthly fees. Carriers in the United States still use tiered pricing structures. But apart from technical and financial factors, the United States has been slow to adopt mobile banking for other reasons.

Perhaps our reluctance to part with monthly paper statements has something to do with our inherent mistrust of the intangible. We've all experienced a virus or some other computer glitch that has left us feeling suspicious of our computers. Some of us naively mistrust online, digital transactions; yet, data tells us that fraud occurs as often, if not more, with paper documents in the physical world. Some of us simply refrain from shopping online and therefore remain ambivalent about going mobile.

In developing countries, there is no such history. By skipping the telephone-based Internet, with all its dial-up problems, its inconsistent Wi-Fi connections, and its malware, these villagers have embraced the mobile phone out of necessity as it provides them with basic services in ways few of us have imagined.

And this is where the lessons learned from hardware hacking, the research presented in this book, can start to pay dividends.

We have seen that layering—adding not subtracting security—has benefits. All locks, whether they are mechanical or virtual, can be defeated. The best we can ever hope for is to slow the criminal down for long enough that someone intervenes or, better, the criminal simply gives up.

We have also seen how the complexity of individual features leads to inherent weaknesses in systems. Feature-rich hotel TV remotes can expose sensitive information about guests, including billing details. And wireless tire-pressure monitor systems, originally designed to protect drivers, can be exploited to attack other computer systems within the modern car. Both of these systems can be compromised because they were not designed with security in mind. Security should be built in first, not bolted on later.

With the mobile phone, we are still at a point at which we can strongly influence the design of future phones; therefore, we should demand, as consumers, better built-in security on new mobile gadgets. Chips should authenticate users and guard against rogue software upgrades. And data should be encrypted.

Unlike David Beckham's luxury car, with its one-click antitheft solution, the mobile gadget should have multiple layers of security. Handset manufacturers are toying with fingerprint sensors; others are considering voice recognition. As we saw in Chapter 6, these solutions alone would not be good enough, but bundled with a PIN or a unique code embedded on a chip inside the handset, they might make the mobile gadget secure enough for financial transactions.

You might wonder why relying on one gadget would be a good idea. For one thing, we're already beholden to our mobile phones. Most people are loath to leave home without their mobile phone, yet leave their keys and wallet behind with some regularity. And most people in the United States upgrade their mobile phone every two years, providing ample opportunity for the handset manufacturers to secure the chip set inside and provide secure software as well. We just need to make it clear to the hardware community that their products can be hacked and can fail just as easily as any software product. We as consumers need to push for this change.

Having secured the mobile gadget, we can add Near Field Communication (NFC), a limited form of RFID with a very low-energy, narrow field of use. NFC has a much shorter range than RFID or Bluetooth. One could swipe an NFC-enabled mobile gadget over an electronic parking meter or transit turnstile and eliminate the need to carry a smart card. For that matter, one could swipe a mobile gadget over a wall plate and gain access to an office on a corporate campus. This would keep criminals from cloning our plastic smart cards by completely removing them from the equation. An electronic wallet might work as long as we remained in control.

John Hering and Kevin Mahaffey, who both worked with Martin Herfurt to extend the range of Bluetooth (discussed in Chapter 3), have created a start-up company in San Francisco that proposes a radical change in how we secure our mobile gadgets. Going beyond just providing antivirus and antimalware software, the Lookout Mobile Security suite provides a number of different protections for the mobile phone, including monitoring the actions of applications installed on the device, such as whether they are leaking information or otherwise spying on us.[11]

In the future, a thief might steal your mobile gadget, but you will be able to prevent him from accessing your data or, for that matter, ever using your phone. Lookout includes something innovative: the ability to report your mobile as stolen, lock it, and, if necessary, erase the data it contains remotely. Because it's connected to the Internet, the mobile gadget will always be in your control, even if you aren't in physical possession of it.

While your carrier can remotely locate or wipe your phone as well, given the general availability of the Internet, it seems logical that you will be able to find an Internet connection faster or maybe use a friend's smart phone to stop a thief within the first few minutes. Simply type in your user name and password, and the thief now has an expensive brick in his hand. This innovation should also begin to change our relationship with personal gadgets, at least mobile phones.

Already enterprises are doing something similar with BlackBerry devices and other mobile devices loaded with sensitive corporate information.[12] Remote backups and remote kill features are becoming common mobile IT practices. But something else is happening, a larger paradigm shift.

The idea of having a strong perimeter (firewalls next to firewalls) is falling out of favor within the security community. With the advent of Instant Messaging, Web mail, and peer-to-peer communication, corporate firewalls are already as porous as Swiss cheese, with employees able to instant-message or e-mail sensitive information. One group of security professionals, the Jericho Forum, advocates (among other things) the responsible removal of the perimeter, and as an alternative, they're focusing more on protecting the data itself.[13] Some enterprises are starting to do that as well.

Google, for example, does not have the traditional electronic firewall perimeter around its campus; instead, it protects each gadget and the data contained within it. Long ago, said Douglas Merrill, former chief information officer (CIO) and vice president of engineering at Google, the company decided to let its employees use new technologies, like smart phones and netbooks, and any operating system they wanted—Microsoft, Apple, Linux—rather than trying to police them. Merrill recognized that Google couldn't effectively protect the network from every new gadget. So the company decided to protect the data, through access rights and encryption, and the gadgets themselves, through security software on the gadget itself; otherwise, the perimeter was left open. This radical rethinking of security has yet to take off, but I think it will be more common in the coming decade.

Merrill used the well-known metaphor of placing sidewalks on a college campus to describe his view of security architecture. Groundskeepers, he said, put down elegant sidewalks and grass. Six months go by, and they begin to notice patches of dead grass where the students actually walk as opposed to using the sidewalks intended for the traffic. Under existing security practices, the groundskeepers would put up metal chains to redirect the students onto the paths. The students, however, would persist in walking where they should not, so the planners would put in planters to discourage such traffic. The solution, he said, is to lay the paths where the students actually walk, then maintain those pathways instead.[14]

The same happens with security in a large enterprise. Companies will try to control employees by restricting Instant Messaging use or by forcing the use of Gmail through a proxy. Citing his own experience as CIO and his frustration with Outlook Exchange, Merrill faulted the classic enterprise software for not being entirely user-friendly. "Employees want better tools at work," said Merrill. "They're trying to make use of the best technology." In his opinion, the best technology can be found in consumer grade—not enterprise grade—software.[15]

As a result, we need to rethink how we secure each gadget—and completely remove its data if necessary. For starters, we should all change our default settings to secure our connections. And this goes back to Merrill's comment that we really should be protecting the devices and the data and not the perimeter.

Accepting that will require a major change in our relationship to new technology—and perhaps even a little technological common sense. We'll need to take apart everything, scrutinize the gadgets we now take for granted, and view with suspicion new gadgets that come our way. As Deviant Ollam reminds us, "Exposing bad security is what protects us all."[16]

ACKNOWLEDGMENTS

I would like to thank B33 and Bunny for making my life very interesting during the time I wrote this book; Tananarive Due, whose lifelong friendship continues to inspire me; Tony Dollar, who first got me thinking about nonfiction over a decade ago, and his wife, Diana, who became my intended audience for this particular book. And for their continuing love and support while I "disappeared" from planet Earth, I thank my family: Garry, Joanne, Michelle, Mike, Ben, Noel, and, of course, my father.

Special thanks to my agent, David Fugate, at LaunchBooks Literary Agency for believing, and to my editor, Tim Sullivan, at Basic Books for "getting it." Also at Basic, I thank my editors, T. J. Kelleher and Adam Khatib; my project manager, Melissa Veronesi; and my copyeditor, Jennifer Kelland.

For access to some of the greatest security researchers on Earth, I thank Ping Look at Black Hat, Bruce Potter at ShmooCon, and Nico Sell at MontaraMountain. And for great pressroom discussions, thanks to my security colleagues Joris Evers, Brian Krebs, Rob Lemos, Declan McCullagh, Bob McMillan, Elinor Mills, Ryan Naraine, and Kim Zetter.

I also want to thank my various editors: Mike Ricciuti, who first thought it was possible for me to work full-time and write this book over an endless series of nights and weekends; Steve Fox, who gave me a helping hand when I needed it most; Robert Strohmeyer and Nick Mediati at PCWorld, who, along with Andy Greenberg at Forbes.com, keep me relevant within the security world; and Brian Livingston and

Tracey Capen at Windows Secrets, who get me out before a different technical audience each month.

For their profound insight, I humbly thank my first readers: Joe Grand, Paul Henry, Benjamin Jun, Kevin Mahaffey, Dave Marcus, Joshua Marpet, Jeff Moss, Chris Paget, and Deviant Ollam. And I thank those I've interviewed for the book over the years: Ian Angell, John Hering, Mikko Hypponen, Eugene Kaspersky, Adam Laurie, Douglas Merrill, Jason Ostrom, Chris Paget, Adrian Pastor, and Howard Schmidt, with sincere apologies to anyone I may have missed. Any published mistakes are mine and mine alone.

My writing music: Y34RZ3R0R3M1X3D, The Slip, and HTDA, and for the final edits, *The Social Network* soundtrack.

NOTES

INTRODUCTION

1. http://news.bbc.co.uk/2/hi/uk_news/wales/south_west/6646331.stm.

2. www.timesonline.co.uk/tol/news/article707216.ece.

3. www.theregister.co.uk/2007/03/26/subaquatic_merc.

4. www.engadget.com/2006/10/10/motorist-has-faith-in-gps-drives-into-sand pile.

5. http://news.bbc.co.uk/2/hi/uk_news/england/london/5172082.stm.

6. www.abiresearch.com/press/1039.

7. http://dev.inversepath.com/rds/cansecwest_2007.pdf.

8. www.masshightech.com/stories/2008/01/14/daily20-Skyhook-powers-Apples -new-location-apps.html.

9. www.tomshardware.com/news/researchers-crack-iphone-s-wi-fi-positioning -system,5201.html.

10. http://thebmxr.googlepages.com/Don_t_Locate_me.pdf.

11. In 2010, Samy Kamkar, author of the 2005 MySpace Samy worm, described how one could steal Media Access Control information from PCs that visited a specially compromised website. See www.bbc.co.uk/news/technology-10850875.

12. Personal interview with Joe Grand, Arlington, Virginia, February 2, 2010. Unless otherwise indicated, all quotes attributed to Grand in this chapter are from this interview.

13. ftp://download.intel.com/museum/Moores_Law/Articles-Press_Releases/ Gordon_Moore_1965_Article.pdf.

14. Personal interview with Deviant Ollam, Arlington, Virginia, February 2, 2010. Unless otherwise indicated, all quotes attributed to Deviant Ollam in this chapter are from this interview.

15. Instant fingerprint analysis isn't 100 percent accurate.

CHAPTER 1

1. www.iaati.org.

2. www.praguepost.com/archivescontent/1529-car-thief-brags-of-prolific-success .html.

3. www.fbi.gov/ucr/prelimsem2009/downloads.htm.

4. www.nicb.org.

5. See "Cracking the key to car immobilisers," *New Scientist*, December 4, 2010, page 21 for more about the general weakness of car immobilizers, and the dramatic rise of auto thefts in Germany in 2009 after a 16-year decline.

6. www.praguepost.com/archivescontent/1529-car-thief-brags-of-prolific-success .html.

7. www.madsci.org/posts/archives/may99/926999832.Eg.r.html.

8. ftp://download.intel.com/museum/Moores_Law/Articles-Press_Releases/ Gordon_Moore_1965_Article.pdf.

9. www.praguepost.com/archivescontent/1529-car-thief-brags-of-prolific-success .html.

10. www.wired.com/wired/archive/14.08/carkey.html.

11. http://webcache.googleusercontent.com/search?q=cache:Lg0qygTm3lcJ: www.msnbc.msn.com/id/4265482/ns/newsweek-newsweek_technology+brad +stone+msnbc+honky&cd=1&hl=en&ct=clnk&gl=us&client=firefox-a.

12. www.vintrack.com/Response%20to%20Wired%20Article.pdf.

13. www.langetech.net/uploads/Response%20to%20Pinch%20My%20Ride% 2010.03.06.pdf.

14. www.wired.com/wired/archive/14.08/carkey.html.

15. www.washingtonpost.com/wp-dyn/content/article/2006/03/13/AR200603 1301822.html.

16. Steven Levy, *Hackers* (New York: Doubleday Books, 1984), 102.

17. www.wired.com/techbiz/people/magazine/17–06/ff_keymaster.

18. http://deviating.net.

19. www.blackhat.com/presentations/bh-europe-08/Deviant_Ollam/White paper/bh-eu-08-deviant_ollam-WP.pdf.

20. For more on cracking the Medeco biaxial lock, see Marc Weber Tobias and Tobias Bluzmanis, *Open in Thirty Seconds: Cracking One of the Most Secure Locks in America* (Sioux Falls, SD: Pine Hill Press, 2008).

21. Personal interview with Deviant Ollam, Arlington, Virginia, February 2, 2010. Unless otherwise indicated, all quotes attributed to Deviant Ollam in this chapter are from this interview.

22. www.wired.com/threatlevel/2008/08/medeco-locks-cr.

23. http://books.google.com/books?id=49IDAAAAMBAJ&pg=PA33&lpg=PA33 &dq=popular+mechanics+kryptonite+tobias&source#v=onepage&q&f=false.

24. www.wired.com/culture/lifestyle/news/2004/09/64987.

25. www.bikeforums.net.

26. http://books.google.com/books?id=49IDAAAAMBAJ&pg=PA33&lpg=PA33 &dq=popular+mechanics+kryptonite+tobias&source#v=onepage&q&f=false.

27. It should be noted that the original tubular lock design was the product of the Chicago Lock Company, an organization that did a very good job of designing and manufacturing the lock. Their product, the ACE Lock, remains the mark of

top quality in this field. However, since the design itself has become public domain, many other vendors are now producing similar tubular locks of much lower quality and with many exploitable weaknesses.

28. www.brickbats.co.uk/D-locks.pdf.

29. www.thesidebar.org/insecurity/?p=447.

30. www.thesidebar.org/insecurity/?p=447.

31. www.slate.com/id/2195862.

32. http://books.google.com/books?id=49IDAAAAMBAJ&pg=PA33&lpg=PA33&dq=popular+mechanics+kryptonite+tobias&source#v=onepage&q&f=false.

33. www.boston.com/business/technology/articles/2004/09/16/cyclists_bike_locks_easy_prey_for_thieves.

34. www.wired.com/techbiz/people/magazine/17–06/ff_keymaster.

35. http://news.scotsman.com/ViewArticle.aspx?articleid=2769825.

36. http://cars.uk.msn.com/news/articles.aspx?cp-documentid=147862479.

37. http://news.bbc.co.uk/2/hi/uk_news/england/2393681.stm.

38. http://sportscustomclassiccars.suite101.com/article.cfm/the_david_beckham_car_collection.

39. http://articles.latimes.com/2007/jul/08/sports/sp-beckham8?pg=2.

40. www.autoblog.com/2006/04/25/celebrity-shocker-beckhams-bimmer-boosted-again.

41. http://news.scotsman.com/ViewArticle.aspx?articleid=2769825.

42. http://news.scotsman.com/davidbeckham/Beckhams-BMW-found-in-Macedonia.2829582.jp.

43. www.nzherald.co.nz/motor-vehicles/news/article.cfm?c_id=65&objectid=10434181.

44. www.autoblog.com/2006/04/25/celebrity-shocker-beckhams-bimmer-boosted-again.

45. http://findarticles.com/p/news-articles/daily-mirror-the-london-uk/mi_8006/is_20060424/beck/ai_n40835831.

46. www.leftlanenews.com/david-beckham-x5-stolen-again.html.

47. www.nytimes.com/2007/04/14/automobiles/14keys.html?_r=2&oref=slogin.

48. www.greenmotor.co.uk/2006/05/keyless-or-clueless-part-2.html.

49. www.community-newspapers.com/archives/campbellreporter/20051130/ca-newsbriefs.shtml.

50. http://sccounty01.co.santa-cruz.ca.us/bds/Govstream/BDSvData/non_legacy/agendas/2007/20070522/PDF/021.pdf.

51. www.cc.gatech.edu/~traynor/f08/slides/lecture15-rfid.pdf.

52. www.carthiefstoppers.com/About-RFIDs-and-the-Texas-Intruments-DST.html.

53. www.nytimes.com/2005/01/28/science/28cnd-key.html?_r=1&pagewanted=1&ei=5094&en=48eb306a45a3b7a0&hp&ex=1106974800&partner=homepage.

54. www.langetech.net/uploads/Response%20to%20Pinch%20My%20Ride%2010.03.06.pdf.

55. http://securityevaluators.com/content/case-studies/tiris/index.jsp.

56. www.theregister.co.uk/2007/08/24/car_cypher_crack.

57. www.sundayherald.com/news/heraldnews/display.var.2174801.0.scientists_crack_security_system_of_millions_of_cars.php.

58. http://blackhat.com/presentations/bh-dc-10/Deviant_Ollam/BlackHat-DC-2010-Deviant%20Ollam-The-Four-Types-of-Lock-wp.pdf.

59. www.wikihow.com/Crack-a-%22Master-Lock%22-Combination-Lock.

60. For more on the craft of lock picking, see http://toool.us.

61. Deviant always describes this fourth category using quotation marks since, he said, no lock should ever officially be considered unpickable in an absolute sense.

62. www.edmunds.com/reviews/list/top10/103271/article.html.

63. www.nicb.org/File Library/Theft and Fraud Prevention/Fact Sheets/Public/layeredanti-theftdevicespublic.pdf.

CHAPTER 2

1. www.thebunker.net/facilities.

2. www.zdnet.com/news/hacking-the-hotel-through-the-tv/143975.

3. www.defcon.org/images/defcon-13/dc13-presentations/DC_13-Major Malfunction.pdf.

4. www.zdnet.com/news/hacking-the-hotel-through-the-tv/143975.

5. www.wired.com/politics/security/news/2005/07/68370.

6. www.zdnet.com/news/hacking-the-hotel-through-the-tv/143975.

7. Personal notes from Black Hat Briefings, Las Vegas, Nevada, July 2007.

8. www.wired.com/politics/security/news/2005/07/68370.

9. www.alcrypto.co.uk/MMIrDA.

10. http://fora.tv/2007/09/27/Implementing_Science_Fiction.

11. Johnny Long, *No-Tech Hacking* (Burlington, MA: Syngress Press, 2008).

12. www.theregister.co.uk/2005/08/22/hotel_hacking_reloaded.

13. www.wired.com/politics/security/news/2005/07/68370.

14. Personal notes from Black Hat Briefings, Las Vegas, Nevada, July 2007.

15. In the summer of 2010, the Stuxnet worm became the first to target SCADA systems: www.computerworld.com/s/article/9179689/Stuxnet_renews_power_grid_security_concerns.

16. http://thomas.loc.gov/cgi-bin/bdquery/z?d110:HR00006:.

17. http://homeland.house.gov/press/index.asp?ID=450.

18. www.cbsnews.com/stories/2009/11/06/60minutes/main5555565.shtml.

19. www.wired.com/threatlevel/2009/11/brazil_blackout.

20. www.nytimes.com/2009/11/12/world/americas/12brazil.html.

21. www.wired.com/threatlevel/2009/11/brazil_blackout.

22. www.wired.com/threatlevel/2009/10/smartgrid.

23. www.edisonfoundation.net.

24. https://media.blackhat.com/bh-us-10/whitepapers/Pollet_Cummins/BlackHat-USA-2010-Pollet-Cummings-RTS-Electricity-for-Free-wp.pdf.

25. https://media.blackhat.com/bh-us-10/whitepapers/Pollet_Cummins/BlackHat-USA-2010-Pollet-Cummings-RTS-Electricity-for-Free-wp.pdf.

26. https://media.blackhat.com/bh-us-10/video/Moyer_Keltner/BlackHat-USA-2010-Moyer-Keltner-Wardriving-the-Smart-Grid.m4v.

27. http://technologyreview.com/computing/25920/page1.

28. www.blackhat.com/presentations/bh-usa-09/FLICK/BHUSA09-Flick-Hacking SmartGrid-SLIDES.pdf.

29. www.blackhat.com/presentations/bh-usa-09/FLICK/BHUSA09-Flick-Hacking SmartGrid-SLIDES.pdf.

30. www.blackhat.com/presentations/bh-usa-09/MDAVIS/BHUSA09-Davis-AMI-SLIDES.pdf.

31. www.internetnews.com/infra/article.php/3831956/Black-Hat-Exposes-Smart-Grid-Security-Risks.htm.

32. www.theregister.co.uk/2009/06/12/smart_grid_security_risks.

33. Personal interview with Howard Schmidt, San Francisco, California, April 20, 2009. Unless otherwise indicated, all quotes attributed to Schmidt in this chapter are from this interview.

34. www.cs.washington.edu/research/security/usenix07devices.pdf.

35. www.businessweek.com/magazine/content/07_02/b4016041.htm.

36. www.infoworld.com/t/data-security/electronic-medical-records-push-must-slow-security-376.

37. Personal interview with Benjamin Jun, San Francisco, California, March 2, 2010. Unless otherwise indicated, all quotes attributed to Jun in this chapter are from this interview.

38. www.youtube.com/watch?v=iE4gTrp_qxw.

39. www.secure-medicine.org/icd-study/icd-study.pdf.

40. www.secure-medicine.org.

41. www.thei3p.org/docs/publications/whitepaper-protecting_global_medical.pdf.

42. www.secure-medicine.org/PervasiveIMDSecurity.pdf.

43. www.nytimes.com/2008/03/12/business/12heart-web.html.

44. www.hacktory.cs.columbia.edu.

45. www.wired.com/threatlevel/2009/10/vulnerable-devices.

46. http://ids.ftw.fm/Home/publications/RouterScan-RAID09-Poster.pdf?attredirects=0.

47. www.nytimes.com/2008/03/12/business/12heart-web.html.

48. www.rfidvirus.org/papers/percom.06.pdf.

49. http://reviews.cnet.com/4520–3513_7–6466679–1.html.

50. www.rfidvirus.org/papers/percom.06.pdf.

51. www.autosec.org/pubs/cars-oakland2010.pdf.

52. For one example, see http://arb.ca.gov/msprog/aftermkt/devices/eo/D-652.pdf.

53. www.autosec.org/pubs/cars-oakland2010.pdf.

54. www.winlab.rutgers.edu/~Gruteser/papers/xu_tpms10.pdf.

55. www.thetechherald.com/article.php/201010/5347/Ford-to-offer-security-with-second-generation-SYNC-system.

56. www.octane.ie/forum/showthread.php?t=177.

57. www.autosec.org/pubs/cars-oakland2010.pdf.

58. L33T is a language used by computer users, particularly gamers, who don't want their work picked up by common Web searches. "Pwn" is L33T language for "own." For more, see www.urbandictionary.com/define.php?term=l33t.

59. www.autosec.org/pubs/cars-oakland2010.pdf.

60. Debora Viana Thompson, Rebecca W. Hamilton, and Roland T. Rust, "Defeating Feature Fatigue," *Harvard Business Review* 84, no. 2 (February 2006): 7.

61. www.schneier.com/crypto-gram-0003.html.

62. www.schneier.com/crypto-gram-0003.html.

CHAPTER 3

1. www.theregister.co.uk/2010/08/02/gsm_cracking.

2. http://events.ccc.de/congress/2009/Fahrplan/events/3654.en.html.

3. www.youtube.com/watch?v=rl5uq7EzVYQ.

4. There is an A5/0, but it offers no encryption.

5. www.securitytube.net/GSM-SRSLY-%28Shmoocon-2010%29-video.aspx.

6. http://tools.ietf.org/html/rfc791.

7. In 2007, Alexander Lash did present a CDMA hack at ToorCon 9; however, it was a way to get a less expensive data plan through Verizon's VCast and not an eavesdropping attack like those on GSM discussed here. See more at http://hackaday .com/2007/10/21/toorcon-9-cdma-unlocking-and-modification.

8. http://thehiphopconsultant.com/2010/01/05/computer-hackers-cracks -secret-codes-to-80-of-the-worlds-cell-phones.

9. http://news.bbc.co.uk/2/hi/8429233.stm.

10. www.nytimes.com/2009/12/29/technology/29hack.html.

11. From Nohl's presentation at Black Hat USA. See http://media.blackhat .com/bh-us-10/whitepapers/Nohl/BlackHat-USA-2010-Nohl-Attacking.Phone .Privacy-wp.pdf.

12. https://media.blackhat.com/bh-us-10/presentations/Grugq/BlackHat-USA -2010-Gurgq-Base-Jumping-slides.pdf.

13. http://blogs.forbes.com/firewall/2010/07/30/cell-phone-snooping-cheaper -and-easier-than-ever.

14. This is mentioned during his presentation. See https://media.blackhat .com/bh-us-10/presentations/Grugq/BlackHat-USA-2010-Gurgq-Base-Jumping -slides.pdf.

15. http://blogs.forbes.com/firewall/2010/07/31/despite-fcc-scare-tactics -researcher-demos-att-eavesdropping.

16. http://blogs.forbes.com/firewall/2010/07/31/despite-fcc-scare-tactics -researcher-demos-att-eavesdropping.

17. http://blogs.forbes.com/firewall/2010/07/31/despite-fcc-scare-tactics -researcher-demos-att-eavesdropping.

18. http://reviews.cnet.com/4520–3513_7–6712994–1.html.

19. The term comes from a 2002 paper called "BaseStation Clone (Evil Twin) Intercept Traffic" from Internet Security Systems, no longer online.

20. http://support.microsoft.com/?kbid=917021.

21. Octets are synonymous with bytes. Thus, a megabyte (MB) is sometimes called a megaoctet (MO).

22. http://standards.ieee.org/regauth/groupmac/tutorial.html.

23. See "Defeating Feature Fatigue," *Harvard Business Review*, 84, no. 2 (February 2006).

24. http://www.computerworld.com/s/article/41904/Packet_Switched_vs._Circuit_Switched_Networks.

25. Personal e-mail from Paul Henry, January 18, 2010.

26. Personal e-mail from Paul Henry, January 18, 2010.

27. www.darkreading.com/insiderthreat/security/app-security/showArticle.jhtml?articleID=219000196.

28. http://sandiego.toorcon.org/index.php?option=com_content&task=view&id=16&Itemid=9.

29. www.gnucitizen.org/blog/hacking-video-surveillance-networks.

30. www.gnucitizen.org/projects/total-surveillance-made-easy-with-voip-phones.

31. www.sans.edu/resources/securitylab/voip_se_asia.php.

32. www.shmoocon.org/2008/presentations/Vigilar_VoIP_Hopper_ShmooCon.ppt.

33. http://news.cnet.com/8301–10789_3–9873864–57.html.

34. http://cansecwest.com/csw07/csw07-miras.pdf.

35. www.fcc.gov/oet/ea/fccid.

36. At the time of the book's publication, the FCC changed its site to require registration.

37. www.howstuffworks.com/cordless-telephone.htm.

38. www.chicagonow.com/blogs/chicago-bar-tender/2009/10/lawsuit-baby-monitor-invades-privacy.html.

39. www.women-inventors.com/Hedy-Lammar.asp.

40. www.wirelessdevnet.com/channels/bluetooth/features/bluetooth.html.

41. Or one hundred meters for some devices.

42. http://electronics.howstuffworks.com/bluetooth2.htm.

43. For more details about Bluetooth pairing, see: www.willhackforsushi.com/presentations/icanhearyounow-sansns2007.pdf

44. For more on Bluejacking, see: http://www.brighthub.com/mobile/symbian-platform/articles/16510.aspx

45. http://trifinite.org/trifinite_stuff_bluesnarf.html.

46. www.wifi-toys.com/wi-fi.php?a=articles&id=91.

47. Personal e-mail from Kevin Mahaffey, San Francisco, California, October 19, 2010.

48. http://loosewire.typepad.com/blog/2004/08/welcome_to_long.html.

49. http://news.zdnet.co.uk/security/0,1000000189,39185480,00.htm.

50. www.youtube.com/watch?v=1c-jzYAH2gw.

51. www.quad-a.org/summit/presentations/PMO%20-%20SMITH,%20RON%20MR.,%20LOGISTICS%20CHIEF,%20PM%20UAS.pdf.

52. http://online.wsj.com/article/SB126102247889095011.html.

53. www.aip.org/history/heisenberg.

54. www.physorg.com/news137253732.html.

55. www.blackhat.com/presentations/bh-usa-08/Bienfang/BH_US_08_Bienfang_Quantum_Key_Distribution.pdf.

56. www.nist.gov/public_affairs/releases/quantumkeys_background.htm.

57. www.nist.gov/public_affairs/techbeat/tb2008_0806.htm.

58. It should be noted that two current implementations of quantum communications have been found to have problems: The underlying idea remains sound, but the way in which these companies have chosen to implement it is suspect. See www.nature.com/nphoton/journal/vaop/ncurrent/full/nphoton.2010.214.html.

CHAPTER 4

1. http://query.nytimes.com/gst/fullpage.html?res=9902E5D9133FF930A25757C0A9619C8B63&sec=&spon=&pagewanted=all.

2. www.nytimes.com/2007/04/13/nyregion/20070413_CORZINE_GRAPHIC.html?_r=1.

3. http://query.nytimes.com/gst/fullpage.html?res=9E03E2DB1E3FF93BA25757C0A9619C8B63&sec=&spon=&pagewanted=all.

4. www.airbagcrash.com.

5. www.nhtsa.dot.gov/cars/rules/rulings/edrnprm4—june1/part2.html.

6. www.nytimes.com/2010/08/11/business/11auto.html.

7. www.nytimes.com/2010/08/11/business/11auto.html?_r=1.

8. On the other hand, Toyota's own investigation did find instances in which the pedals did not work properly, and because the company did not report the mechanical problems when they were first discovered, Toyota was fined $16.4 million.

9. www.nytimes.com/2010/08/11/business/11auto.html?_r=1.

10. www.nhtsa.dot.gov/cars/rules/rulings/edrnprm4—june1/part2.html.

11. Thomas Kowalick, *Fatal Exit: The Automotive Black Box Debate* (Hoboken, NJ: John Wiley & Sons and IEEE Press, 2004).

12. www.csmonitor.com/2004/1227/p13s01-wmgn.html.

13. Kowalick, *Fatal Exit*.

14. www.eff.org/deeplinks/2009/07/pay-you-drive-black-.

15. www.motorists.org/edr/home/maine-governor-disputes-black-box-data.

16. www.wired.com/politics/security/news/2005/06/67952.

17. www.motorists.org/black-boxes/failure.

18. www.motorists.org/black-boxes/black-box-info-packet.pdf.

19. Patrick Mueller, "Every Time You Brake, Every Turn You Make—I'll Be Watching You: Protecting Driver Privacy in Event Data Recorder Information," *Wisconsin Law Review* (2006): 135. Available at Social Science Research Network, http://papers.ssrn.com/sol3/papers.cfm?abstract_id=878317.

20. www.nytimes.com/2010/07/22/business/global/22blackbox.html?hp.

21. www.nytimes.com/2010/07/22/business/global/22blackbox.html?hp.

22. www.pcmag.com/article2/0,2817,2346045,00.asp.

23. www.wired.com/autopia/2009/11/mercedes-mbrace-telematics.

24. In 2010, OnStar released an iPod application: http://itunes.apple.com/us/app/onstar-mylink/id393584149?mt=8

25. www.autoweek.com/apps/pbcs.dll/article?AID=/20041108/FREE/41108 0714&SearchID=73279948089928.

26. www.wired.com/wired/archive/7.07/gm.html.

27. http://gpsobsessed.com/gm-onstar-now-instantly-sends-accident-location -coordinates-to-911.

28. http://gpsobsessed.com/onstar-adds-remote-ignition-block-to-thwart-car -thieves.

29. www.onstar.com/us_english/jsp/privacy_policy.jsp.

30. www.pcworld.com/article/127823/the_13_most_embarrassing_web _moments.html.

31. www.pcworld.com/article/205296/what_your_digital_photos_reveal _about_you.html.

32. www.mayhemiclabs.com/files/Locational%20Privacy%20and%20Whole sale%20Surveillance.pdf.

33. http://code.google.com/p/icanstalku/source/browse/trunk/stalk.pl.

34. www.networkworld.com/news/2010/091310-what-your-digital-photos -reveal.html?hpg1=bn.

35. www.computerworld.com/s/article/9127717/Survey_40_of_hard_drives _bought_on_eBay_hold_personal_corporate_data?taxonomyId=19&intsrc=kc _top&taxonomyName=storage.

36. www.zdnetasia.com/news/hardware/0,39042972,38025936,00.htm.

37. www.slate.com/id/2168757.

38. http://simson.net/clips/2003/2003.CSO.04.Hard_disk_risk.htm.

39. www.informit.com/guides/content.aspx?g=security&seqNum=234.

40. http://reviews.cnet.com/4531–10921_7–6719504.html (accessed February 28, 2010).

41. www.msnbc.msn.com/id/17597505/ns/technology_and_science-security.

42. http://homelandsecuritynewswire.com/bay-areas-fastrak-road-tolls-easy -hack.

43. http://bata.mtc.ca.gov/faq.htm.

44. www.highbeam.com/doc/1P2–7456514.html.

45. www.recordonline.com/apps/pbcs.dll/article?AID=/20090419/NEWS/9041 90316/-1/NEWS67.

46. www.highbeam.com/doc/1P2–7456514.html.

47. www.highbeam.com/doc/1P2–7456514.html.

48. www.msnbc.msn.com/id/20216302.

49. http://blog.nj.com/ledgerupdates/mcguire_murder_trial/index_3.html.

50. Matzan's story is chronicled at www.thejemreport.com/content/view/ 385/74.

51. http://tech2.nytimes.com/mem/technology/techreview.html?res=980D E5DE1038F932A15750C0A9649C8B63.

52. http://sip-trunking.tmcnet.com/topics/security/articles/55192-interview -intellicheck-mobilisa-defense-id-system.htm.

53. www.nytimes.com/2005/03/06/fashion/06fake.html.

54. http://tech2.nytimes.com/mem/technology/techreview.html?res=980D E5DE1038F932A15750C0A9649C8B63.

55. www.wired.com/politics/law/news/2000/01/33611.

56. http://articles.latimes.com/1989–07–23/news/mn-406_1_rebecca-schaeffer.

57. http://articles.latimes.com/1989–07–23/news/mn-406_1_rebecca-schaeffer.

58. www.imdb.com/name/nm0757854.

59. www.smartvoter.org/1998jun/ca/or/vote/watson_j/paper1.html.

60. http://thomas.loc.gov/cgi-bin/bdquery/z?d103:s.01589:.

61. http://thomas.loc.gov/cgi-bin/bdquery/z?d103:h.r.03365:.

62. www.congress.gov/cgi-bin/bdquery/z?d103:HR03355:.

63. www.deadiversion.usdoj.gov/meth/index.html.

64. www.docstoc.com/docs/1085827/General-Information-regarding-the -Combat-Methamphetamine-Epidemic-Act-of-2005.

65. www.pharmitas.com/about/Bartell%20Drug%20Company.pdf.

66. See www.nybooks.com/articles/23231 for a review of Matthew M. Aid's *The Secret Sentry: The Untold History of the National Security Agency* (New York: Bloomsbury Press, 2009).

67. www.nybooks.com/articles/23231.

68. An account of the Dutch experience is described in Viktor Mayer-Schonberger, *Delete* (Princeton, NJ: Princeton University Press, 2009).

CHAPTER 5

1. Personal interview with Joe Grand, Arlington, Virginia, February 2, 2010. Unless otherwise indicated, all quotes attributed to Grand in this chapter are from this interview.

2. Appelbaum's joke was made during the presentation at Black Hat and recorded in personal notes taken at the event. See www.blackhat.com/presentations/ bh-usa-09/GRAND/BHUSA09-Grand-ParkingMeter-SLIDES.pdf.

3. www.chicagobreakingnews.com/2009/12/laz-parking-chicago-meter -electronic.html.

4. www.nytimes.com/2009/12/25/us/25cncpulse2.html?_r=1.

5. www.nytimes.com/2009/05/30/us/30parking.html.

6. http://cbs2chicago.com/local/Levin.pay.parking.2.1027406.html.

7. http://blogs.suntimes.com/marin/2009/03/chicago_parking_meter_rebellio .html.

8. http://theexpiredmeter.com/?p=3389.

9. www.nydailynews.com/archives/news/2001/02/14/2001–02–14_tv_remotes _zap_meters_city_f.html.

10. www.nytimes.com/2006/08/31/business/31sbiz.html.

11. Shortly after the Clipper card went into service in 2010, San Francisco residents found various ways to cheat the fares: www.sfgate.com/cgi-bin/blogs/transportation/detail?entry_id=76657

12. http://video.google.com/videoplay?docid=4252367680974396650&hl=en#.

13. www.blackhat.com/presentations/bh-usa-08/Nohl/BH_US_08_Nohl_Mifare.pdf.

14. www.nxp.com/news/content/file_1569.html.

15. www.ru.nl/ds/research/rfid.

16. www.youtube.com/watch?v=NW3RGbQTLhE.

17. http://news.cnet.com/8301–10789_3–9978486–57.html.

18. http://news.bbc.co.uk/2/hi/7516869.stm.

19. www.securitysystemsnewseurope.com/?p=article&id=se200908Tjypkc.

20. http://news.bbc.co.uk/2/hi/7516869.stm.

21. www.popularmechanics.com/technology/industry/4278892.html.

22. www.defcon.org/images/defcon-16/dc16-presentations/anderson-ryan-chiesa/1-complaint.pdf.

23. www.law.cornell.edu/uscode/18/usc_sec_18_00001030——000-.html.

24. www.zdnet.com/blog/security/hid-denies-rfid-demo-threat-hackers-worry/103.

25. Cisco not only removed content from the bound conference books but also had to reburn the conference CDs without the slides—all within twenty-four hours of the start of Black Hat USA 2005. See http://news.cnet.com/Cisco-details-controversial-router-flaw/2100–1002_3–5810669.html.

26. http://tech.mit.edu/V128/N30/subway/Defcon_Presentation.pdf.

27. One of Adam Laurie's websites is http://rfidiot.org.

28. www.blackhat.com/presentations/bh-europe-07/Laurie/Presentation/bh-eu-07-laurie.pdf.

29. www.wired.com/threatlevel/2009/08/fed-rfid.

30. I was at the conference in Washington, DC, and various other sources can confirm this information.

31. www.wired.com/threatlevel/2009/08/fed-rfid.

32. www.wired.com/threatlevel/2009/08/fed-rfid.

33. www.youtube.com/watch?v=NW3RGbQTLhE.

34. www.prisonplanet.com/articles/april2004/040704bajabeachclub.htm.

35. www.zdnet.com/blog/healthcare/verichip-going-away/1027.

36. www.ama-assn.org/ama1/pub/upload/mm/467/ceja5a07.doc.

37. www.antichips.com/faq/html/faq-section03.html#toc19.

38. http://blogs.reuters.com/blog/2006/07/22/high-tech-cloning.

39. http://cq.cx/verichip.pl.

40. www.wired.com/threatlevel/2008/10/rfid-anti-skimm.

41. www.blackhat.com/presentations/bh-europe-07/Laurie/Presentation/bh-eu-07-laurie.pdf.

42. http://news.cnet.com/8301–10789_3–9875961–57.html.

43. http://news.cnet.com/8301–10789_3–9875961–57.html.

44. www.epcglobalinc.org/standards/uhfc1g2.

45. www.pcworld.com/article/129625/uk_biometric_passports_not_secure.html.

46. www.dailymail.co.uk/news/article-440069/Safest-passport-fit-purpose.html.

47. http://travel.state.gov/passport/passport_2788.html.

48. http://hasbrouck.org/documents/ICAO9303-pt1-vol2.pdf.

49. www.youtube.com/watch?v=-XXaqraF7pI&p=AC4AA6957C894CEE& playnext=1&index=6.

50. www.youtube.com/watch?v=-XXaqraF7pI&p=AC4AA6957C894CEE& playnext=1&index=6.

51. www.rsa.com/rsalabs/staff/bios/ajuels/publications/EPC_RFID/Gen2 authentication—22Oct08a.pdf.

52. www.usatoday.com/news/nation/2009–05–25-licenseinside_N.htm.

53. www.washingtonpost.com/wp-dyn/content/article/2009/06/13/AR20090 61302036.html.

54. www.usatoday.com/news/nation/2009–05–25-licenseinside_N.htm.

55. www.washingtonpost.com/wp-dyn/content/article/2009/06/13/AR20090 61302036.html.

56. www.rsa.com/rsalabs/staff/bios/ajuels/publications/EPC_RFID/Gen2 authentication—22Oct08a.pdf.

57. www.bnet.com/blog/gadget-guy/walmart-rfid-clothing-tags-create-a-slippery -privacy-slope/458?tag=mantle_skin;content.

58. http://online.wsj.com/article/NA_WSJ_PUB:SB10001424052748704421304 575383213061198090.html#printMode.

59. http://webmonkey.wired.com/wired/archive/12.07/shoppers.html.

60. www.towersnerd.com/articles/item.php?id=95.

61. www.rfidjournal.com/blog/entry/7809.

62. http://itri.uark.edu/rfid.asp.

63. www.celdi.ineg.uark.edu/Conferences/Spring%202006%20Conference/myths %20paper.pdf.

64. www.celdi.ineg.uark.edu/Conferences/Spring%202006%20Conference/myths %20paper.pdf.

65. www.patents.com/identification-tracking-persons-using-rfid-tagged-items -store-environments-7076441.html.

66. www.theregister.co.uk/2010/08/02/long_range_rfid.

67. www.tombom.co.uk/extreme_rfid.pdf.

68. www.tombom.co.uk/extreme_rfid.pdf.

69. www.youtube.com/watch?v=9isKnDiJNPk.

70. This and other scenarios are described in Paget's white paper, "Extreme-Range RFID tracking" (presented at Black Hat USA 2010, Las Vegas, Nevada, July 28–29, 2010). See www.tombom.co.uk/extreme_rfid.pdf.

71. The Fourth Amendment guarantees protection from illegal searches and seizures.

72. www.tombom.co.uk/extreme_rfid.pdf.

73. www.youtube.com/watch?v=9isKnDiJNPk.

74. http://travel.state.gov/travel/cbpmc/cbpmc_2223.html.

75. www.youtube.com/watch?v=9isKnDiJNPk.

CHAPTER 6

1. http://news.bbc.co.uk/2/hi/europe/3504912.stm provides a timeline for the attack.

2. The *New York Times* has several good articles about the Mayfield case at http://topics.nytimes.com/topics/reference/timestopics/people/m/brandon _mayfield/index.html.

3. www.fbi.gov/pressrel/pressrel04/mayfield052404.htm.

4. www.fbi.gov/pressrel/pressrel06/mayfield010606.htm.

5. http://news.bbc.co.uk/2/hi/europe/7766217.stm.

6. www.blackhat.com/presentations/bh-usa-08/Angell/BH_US_Angell_Keynote _Complexity.pdf.

7. Personal interview with Joshua Marpet, Arlington, Virginia, February 3, 2010. Unless otherwise indicated, all quotes attributed to Marpet in this chapter are from this interview.

8. www.newscientist.com/article/dn4611.

9. www.newscientist.com/article/dn4611.

10. www.newscientist.com/article/dn7983.

11. www.popularmechanics.com/technology/military_law/4325774.html.

12. www.nytimes.com/2009/02/05/us/05forensics.html?_r=2&scp=6&sq=DNA &st=cse.

13. www8.nationalacademies.org/onpinews/newsitem.aspx?RecordID=12589.

14. http://judiciary.senate.gov/pdf/09–03–18EdwardsTestimony.pdf.

15. www.youtube.com/watch?v=LA4Xx5Noxyo.

16. http://web.mit.edu/6.857/OldStuff/Fall03/ref/gummy-slides.pdf.

17. http://web.mit.edu/6.857/OldStuff/Fall03/ref/gummy-slides.pdf.

18. www.washingtonpost.com/wp-dyn/content/article/2008/02/11/AR200802 1102786.html.

19. www.theregister.co.uk/2008/03/30/german_interior_minister_fingerprint _appropriated.

20. www.engadget.com/2006/09/01/walt-disney-world-to-start-fingerprinting -everyone.

21. http://newsinitiative.org/story/2006/09/01/walt_disney_world_the _governments.

22. http://newsinitiative.org/story/2006/09/01/walt_disney_world_the _governments.

23. http://portal.acm.org/citation.cfm?id=1263434.

24. http://ftp.rta.nato.int/public/PubFullText/RTO/MP/RTO-MP-IST-044/ MP-IST-044–23.pps.

25. http://news.bbc.co.uk/2/hi/asia-pacific/4396831.stm.

26. www.highbeam.com/doc/1P1–106904746.html.

27. www.sepiamutiny.com/sepia/archives/001285.html.

28. http://news.bbc.co.uk/2/hi/asia-pacific/4396831.stm.

29. www.atmmarketplace.com/article.php?id=9738.

30. www.scientificamerican.com/article.cfm?id=palm-reading-devices.

31. K. Shimizu et al., "Noninvasive Measurement of Physiological Functions in a Living Body by Transillumination," *Proceedings of IEEE Instrument and Measurement Technology Conference '94* 2 (May 1994): 982–985, doi: 10.1109/IMTC.1994.351941.

32. www.docstoc.com/docs/7098475/Vein-pattern-recognition.

33. http://newsinitiative.org/story/2006/09/01/walt_disney_world_the _governments.

34. http://articles.cnn.com/2009–06–23/travel/clear.airport.terminated_1 _security-lanes-airport-security-reagan-national-airport?_s=PM:TRAVEL.

35. www.cl.cam.ac.uk/~jgd1000/irisrecog.pdf.

36. http://news.bbc.co.uk/2/hi/uk_news/england/london/4792206.stm.

37. www.blackhat.com/presentations/bh-dc-08/Franken/Presentation/bh-dc -08-franken.pdf.

38. www.nfl.com/superbowl/history/recap/sbxxxv.

39. www.wired.com/politics/law/news/2001/02/41571.

40. Personal interview with Dr. Angell, Black Hat USA 2008. Unless otherwise indicated, all quotes attributed to Angell in this chapter are from this interview.

41. http://health.howstuffworks.com/mental-health/human-nature/happiness/ muscles-smile.htm.

42. www.nlm.nih.gov/visibleproofs/galleries/biographies/bertillon.html.

43. www.blackhat.com/html/bh-dc-10/bh-dc-10-briefings.html#Marpet.

44. www.duitslandinstituut.nl/binaries/dia/Home/2009/camerasurveillance andfacerecognition_veldhuis_aang.pdf.

45. www.computerworlduk.com/technology/security-products/authentication/ news/index.cfm?newsid=13459.

46. www.govtech.com/gt/374147.

47. http://query.nytimes.com/gst/fullpage.html?res=9907E6D9143EF934A2575 1C0A9619C8B63&sec=&spon=&pagewanted=1.

48. www.washingtonpost.com/wp-dyn/content/article/2009/05/27/AR20090 52703627.html.

49. www.usatoday.com/news/nation/2009–05–25-licenses_N.htm?se=yahoorefer.

50. http://ianangell.blogspot.com/2008/04/fallacy-of-residual-category.html.

51. www.timesonline.co.uk/tol/comment/columnists/guest_contributors/ article3499317.ece.

52. http://homelandsecuritynewswire.com/worries-about-iraqs-biometric -database.

53. http://homelandsecuritynewswire.com/fbi-takes-biometrics-database -proposal-uk.

54. www.wired.com/politics/security/news/2005/09/68973.

55. www.wired.com/politics/security/news/2005/09/68973, and from a personal interview with Dr. Angell.

56. www.timesonline.co.uk/tol/comment/columnists/guest_contributors/ article3499317.ece.

57. www.law.northwestern.edu/wrongfulconvictions.

58. www.law.northwestern.edu/wrongfulconvictions/exonerations/mdBloods worthSummary.html.

59. http://web.archive.org/web/20080202233509/www.courttv.com/trials/ojsimpson/weekly/16.html.

60. http://ianangell.blogspot.com/2008/04/fallacy-of-residual-category.html.

CHAPTER 7

1. Pentland's website (web.media.mit.edu/~sandy) has since been updated, but it once read, "We are in the midst of an explosion of information about people and their behavior, but most of it is noise. Reality Mining sifts through this noise to discover the 'honest signals' hidden within: subtle patterns that reliably reveal our relationships with other people, and accurately predict our future behavior. Honest signals offer an unmatched window into our financial, cultural, and organizational health. By understanding these subtle patterns we can better understand ourselves, and begin to create true collective intelligences."

2. http://hbr.org/web/2009/hbr-list/how-social-networks-work-best.

3. Much of Pentland's research is described fully in Alex ("Sandy") Pentland, *Honest Signals: How They Shape Our World* (Cambridge, MA: MIT Press, 2008).

4. More details about the initial Reality Mining project at MIT can be found at http://reality.media.mit.edu.

5. Nathan Eagle, Alex S. Pentland, and David Lazer, "Inferring Social Network Structure Using Mobile Phone Data," in *Social Computing, Behavioral Modeling, and Prediction* (2008): 79–88, www.socialsciences.cornell.edu/0508/science report_formatted_10.12.pdf.

6. Nathan Eagle and Alex S. Pentland, "Reality Mining: Sensing Complex Social Systems," *Personal and Ubiquitous Computing* 10, no. 4 (2006): 255–268, http://reality.media.mit.edu/pdfs/realitymining.pdf.

7. Eagle and Pentland, "Reality Mining."

8. www.socialsciences.cornell.edu/0508/sciencereport_formatted_10.12.pdf.

9. http://bits.blogs.nytimes.com/2010/02/09/foursquare-inks-a-deal-with-zagat.

10. www.pcworld.com/article/203819/how_to_use_facebook_places.html.

11.www.sfgate.com/cgi-bin/article.cgi?f=/g/a/2010/09/10/businessinsider-burglars-already-using-facebook-places-to-learn-when-its-safe-to-rob-your-home-2010-9.DTL.

12. www.nytimes.com/2008/11/30/business/30privacy.html?_r=1.

13. A video of a lecture given by Pentland at Yahoo! in Santa Clara, California, is available at http://labs.yahoo.com/event/205.

14. http://us.macmillan.com/adamstongue.

15. Malcolm Gladwell, *Blink* (New York: Little Brown & Company, 2005), 21.

16. www.stanford.edu/~bailenso/papers/Digital%20Chameleons,%20in%20press.pdf.

17. www.youtube.com/watch?v=tfECX8VzkIQ.

18. This study is referenced in Pentland, *Honest Signals*.

19. RFID tags from tollbooth transponders have also been used, but they require that readers be placed along the highway. Using cellular data is more universal.

20. www.aip.org/dbis/IEEE/stories/15139.html.

21. www.msnbc.msn.com/id/9698139.

22. www.pcworld.com/article/201605/bank_lost_your_account_data_heres_what_to_do.html.

23. http://events.ccc.de/congress/2008/Fahrplan/attachments/1262_25c3-locating-mobile-phones.pdf.

24. http://events.ccc.de/congress/2008/Fahrplan/attachments/1262_25c3-locating-mobile-phones.pdf.

25. http://thecarmensandiegoproject.com.

26. http://media.blackhat.com/bh-us-10/whitepapers/Bailey_DePetrillo/BlackHat-USA-2010-Bailey-DePetrillo-The-Carmen-Sandiego-Project-wp.pdf.

27. http://blogs.forbes.com/firewall/2010/07/30/cell-phone-snooping-cheaper-and-easier-than-ever/?boxes=Homepagechannels.

28. A full account appears at http://news.ufl.edu/2009/12/16/malaria.

29. http://hd.media.mit.edu/RWJF-Reality-Mining-summary.pdf.

30. http://hd.media.mit.edu/RWJF-Reality-Mining-summary.pdf.

31. http://hd.media.mit.edu/RWJF-Reality-Mining-summary.pdf.

32. http://hd.media.mit.edu/RWJF-Reality-Mining-summary.pdf.

33. www.aahsa.org/section.aspx?id=4672.

34. www.nytimes.com/2008/05/25/us/25aging.html.

35. www.scientificamerican.com/blog/post.cfm?id=ted-med-bringing-medicine-home-for-2009–10–29.

36. www.aahsa.org/section.aspx?id=4672.

37. www.scientificamerican.com/blog/post.cfm?id=ted-med-bringing-medicine-home-for-2009–10–29.

38. www.cse.wustl.edu/~jain/cse574–08/ftp/ban/index.html.

39. www.zigbee.org.

40. www.trilcentre.org.

41. www.biomobius.org.

42. http://abclocal.go.com/kgo/story?section=news/drive_to_discover&id=6214464.

43. www.scientificamerican.com/blog/post.cfm?id=ted-med-bringing-medicine-home-for-2009–10–29.

44. www.mobihealth.org/html/mainframe.html.

45. www.continuaalliance.org.

46. www.continuaalliance.org/products/design-guidelines.html.

47. www.nonin.com/documents/2500%20PalmSAT%20Brochure.pdf.

48. www.continuaalliance.org/products/certified-products.html.

49. http://aahsa.org/cast.aspx.

50. http://aahsa.org/cast.aspx.

51. www.ryanfarley.net/RyanFarley.net/Home.html.

52. http://blogs.abcnews.com/theblotter/2006/12/can_you_hear_me.html.

53. www.katzjustice.com/tomero.pdf.

54. www.katzjustice.com/blog2/serendipity/archives/133-When-are-attorney-client-conversations-safe-from-eavesdropping.html.

55. www.technologyreview.com/Infotech/19968/page1.

56. www.nytimes.com/2008/11/30/business/30privacy.html?_r=1&partner=rss&emc=rss.

57. www.highbeam.com/doc/1P2–13415518.html.

58. www.boston.com/bostonglobe/ideas/articles/2008/03/02/qa_with_alex_pentland.

59. http://hd.media.mit.edu/wef_globalit.pdf, page 6.

60. www.youtube.com/watch?v=Dw3h-rae3uo.

CONCLUSION

1. A full account of Bekowe Skhakhane's mobile banking can be found at www.nytimes.com/2005/08/25/international/africa/25africa.html?_r=1&oref=slogin&pagewanted.

2. www.wizzit.co.za.

3. www.wizzit.co.za/media/wavinghand.pdf.

4. www.safaricom.co.ke/index.php?id=745.

5. www-wds.worldbank.org/external/default/WDSContentServer/IW3P/IB/2008/06/17/000158349_20080617102123/Rendered/PDF/wps4647.pdf.

6. http://blog.mobilephonebanking.rbap.org/index.php/category/news.

7. www.cellular-news.com/story/21391.php.

8. http://blog.foreignpolicy.com/posts/2007/01/17/how_banking_on_a_mobile_phone_can_help_the_poor.

9. One way to use SMS for banking is not to transmit sensitive data; thus, such banking systems are useful mainly for checking account balances and not for paying bills or making transactions.

10. Microsoft discussed its vision for mobile kiosks with blood pressure cuffs and medicine at PDC 2008. See http://channel9.msdn.com/blogs/pdc2008.

11. www.mylookout.com.

12. www.pcworld.com/businesscenter/article/160164/fight_malware_on_the_smartphone.html.

13. www.opengroup.org/jericho/index.htm.

14. www.pcworld.com/businesscenter/article/169356/former_google_vp_suggests_userbased_security_at_black_hat.html.

15. www.pcworld.com/businesscenter/article/169356/former_google_vp_suggests_userbased_security_at_black_hat.html.

16. www.defcon.org/html/links/dc-archives/dc-13-archive.html#Deviant.

INDEX